AIRCRAFT IGNITION AND ELECTRICAL POWER SYSTEMS

AF271220

JEPPESEN.
Sanderson Training Products

Library of Congress Cataloging-in-publication number: 92-25787

JS312656B

Table Of Contents

Preface

This book on *Aircraft Ignition and Electrical Systems* is one of a series of specialized study guides prepared for aviation maintenance personnel, to be used with a corresponding video tape, Aircraft Ignition Systems (Order Number EA-IGS).

This series is part of a programmed learning course developed and produced by IAP, Inc., one of the largest suppliers of aviation maintenance training materials in the world. This program is part of a continuing effort to improve the quality of education for aviation mechanics throughout the world.

The purpose of each IAP, Inc. training series is to provide basic information on the operation and principles of the various aircraft systems and their components.

Specific information on detailed operation procedures should be obtained from the manufacturer through his appropriate maintenance manuals, and followed in detail for the best results.

This particular manual on *Aircraft Ignition and Electrical Systems* includes a series of carefully prepared questions and answers to emphasize key elements of the study, and to encourage you to continually test yourself for accuracy and retention as you use this book. A multiple choice final examination is included to allow you to test your comprehension of the total material.

For best results, the video tape should be reviewed first, either in the classroom under the direction of an experienced instructor, or by individual study; then this material should be reinforced with that included in this text.

Acknowledgements

The validity of any program such as this is enhanced immeasurably by the cooperation shown IAP, Inc. by recognized experts in the field, and by the willingness of the various manufacturers to share their literature and answer countless questions in the preparation of these programs.

We would like to mention, especially, our appreciation for help given us by:

AC Spark Plug Div. of General Motors Corp.

The Bendix Corp., Electrical Components Div.

Champion Spark Plug Co.

Slick Electro, Inc.

Ward Aero, Inc.

If you have any questions or comments regarding this manual, or any of the many other textbooks offered by IAP, Inc., simply contact: Sales Department, IAP, Inc.; Mailing Address: P.O. Box 10000, Casper, WY 82602-1000; Shipping Address: 7383 6WN Road, Casper, WY 82604-1835; or call toll free: (800) 443-9250; in Wyoming/outside continental USA, call: (307) 266-3838.

Introduction

The history of aviation has been a saga of technical developments. Da Vinci envisioned flying in a mechanical device, and since as early as the eighteenth century, men have been aloft in hot air balloons. Wings were spread with gliders, but it took the development of a powerplant to make flying more than just a fun way of breaking our necks.

Steam engines were tried, but they proved too heavy for practicality, and it was not until LeNoir built his internal combustion engine in France in 1860 that a practical approach to airborne power was possible. His engine required a combustible material *inside* the cylinder, and a means of igniting it so it could expand.

Now we are familiar with the evolution the engines themselves have gone through to become the dependable aircraft engine of today. There have been two-cycle, four-cycle, and even experimental engines with other than the familiar suck, squeeze, bang, and blow cycle of operation. Engines have been cooled with water, glycol, and air; cylinders have been arranged in single rows, Vs, fans, Ws, Xs; some have had one crankshaft, some two, and some no crankshaft at all. There have been cam engines and radial engines with one, two, and even four banks of cylinders, all the way up to the gargantuan thirty-six cylinder, liquid-cooled, four-row radial Lycoming R-7770 engine. And lest we forget, even the whirling dervish of World War I, the rotary, popular on both sides of the lines—an engine with the pistons attached to the airframe by a slipper, and the cylinders spinning merrily along with the propeller.

Powerplant configurations and their details have changed—that's the name of the game. But the principles stay the same. Induce a combustible charge, and light it off. This is where the ignition system comes into the limelight. We must have a system that will consistently produce a hot spark at the absolute correct position of the piston—and must do it *every time*. It must be independent of every system in the airplane and, naturally, as a part of a flying machine it must be lightweight.

One of the early approaches to an ignition system was the make-and-break system in which an igniter consisting of a pivoted, insulated contact was installed inside the cylinder and connected through a kick coil to a battery.

The engine, along with its ignition system, made flight possible; but it was only when an electrical system of high output evolved that the airplane itself became really practical. Now we can have ample power for lights, for radio communications and navigation, comfort control systems, and for electrically-operated landing gear and flaps. The present day executive aircraft has an electrical generation and distributing system as complex as that found in the largest airliners of less than a couple of decades ago.

The A & P technician of today must be familiar with all of the details of ignition and electrical systems. Here we assume that you are familiar with the principles covered, but will review the basics of magnetism and voltage generation.

SECTION 1: IGNITION SYSTEMS

Chapter I
Voltage Production

Electrical ignition requires a source of electrical energy. So at this time it is well to review the principles involved in the generation of electrical energy by electromagnetic methods.

Magnetism is one of the true mysteries of nature. While we know very little about the phenomenon of lines of magnetic flux, we do know much about their behavior. If a bar magnet is bent into the form of a horseshoe, lines of magnetic flux will pass between the north and south poles. When a conductor passes between these poles, an electrical current will be caused to flow in the conductor, because of its cutting across the lines of flux. The intensity of the current is determined by the *rate* at which the lines of flux are cut. The more lines cut in an interval of time, the more current.

Increasing the number of lines of flux, increasing the number of turns in the conductor, or increasing the speed at which the conductor moves, will all increase the current.

Lines of flux exist not only between the poles of a permanent magnet, but they surround any conductor carrying current. In Fig. 1-1A, when the conductor is moved out from between the poles, lines of flux encircle the conductor in the direction shown by the arrow. If the intensity of the current changes, the magnetic field will expand and collapse, and if a second conductor is placed so the magnetic field of the first cuts across it (Fig. 1-1B), lines of flux will encircle it, and an induced current will flow.

The induced current also causes a magnetic field which, according to Lenz's law, opposes the field generating it. The voltage, which causes this induced current to flow, is determined by the rate at which the lines of flux are cut. And this in turn is determined by the intensity of the field, the frequency with which the field expands and collapses, and the ratio of the turns in the primary to those in the secondary. The primary winding carries the source current and the secondary the induced current.

**Figure 1-1.
Electromagnetic
generation.**

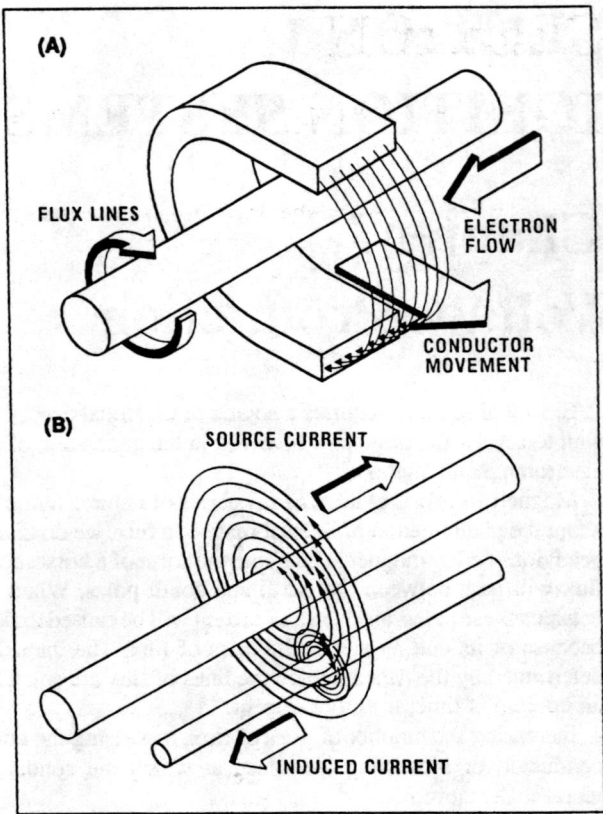

(A)

FLUX LINES

ELECTRON
FLOW

CONDUCTOR
MOVEMENT

(B)

SOURCE CURRENT

INDUCED CURRENT

QUESTIONS:

1. *What happens in a conductor when it cuts across lines of magnetic flux?*
2. *What determines the amount of current that will be generated in a wire moving in a magnetic field?*
3. *Does induced current aid or oppose the current that caused it?*

Chapter II
Battery Ignition System

One of the earliest approaches to an ignition system was the make-and-break system in which an igniter, consisting of a pivoted, insulated contact, was installed inside the cylinder and connected through a kick coil to a battery (Fig. 2-1).

This insulated point made contact with a grounded point and when they were together, current flowed through them to ground. A cam, operated from the crankshaft, opened these points whenever ignition was required. As the points opened, an arc was drawn across them similar to that produced in arc welding.

While current was flowing through the contacts, a magnetic field was built up around the kick coil, and as the points opened, the current tried to decrease. But the energy in the field tried to sustain it, so there was a sustained arc across the points which provided ignition. Naturally, this system had problems, so better things had to be developed.

Aircraft engines now almost universally use magneto ignition, but a few of the early engines used a battery system for starting. Since the principles of the two systems are similar, let's first analyze a battery ignition system (Fig. 2-2).

When the ignition switch is closed, current flows from the battery through the primary winding of the coil and the breaker points to ground. As this current flows, lines of flux build up around each of the turns of the primary and cut across the secondary turns, but since this buildup is rather gradual there will be no appreciable voltage induced into the secondary. When the cam opens the breaker points, the current in the primary abruptly stops flowing, and the magnetic field which has been held out by this current suddenly collapses.

The primary coil is made up of several hundred turns of relatively heavy wire, while the secondary winding has thousands of turns of very small wire. The rapid collapse of the flux from the turns in the primary cutting across the many

Figure 2-1.
Make-and-break
ignition system.

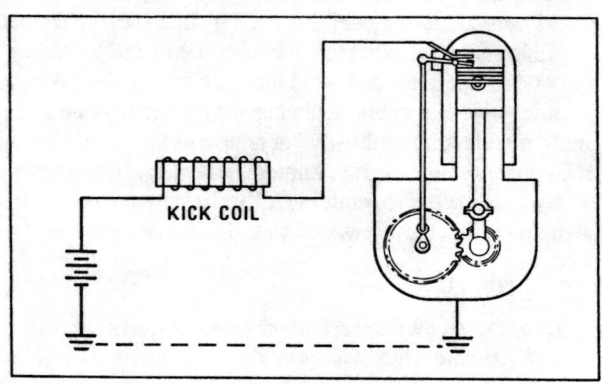

KICK COIL

3

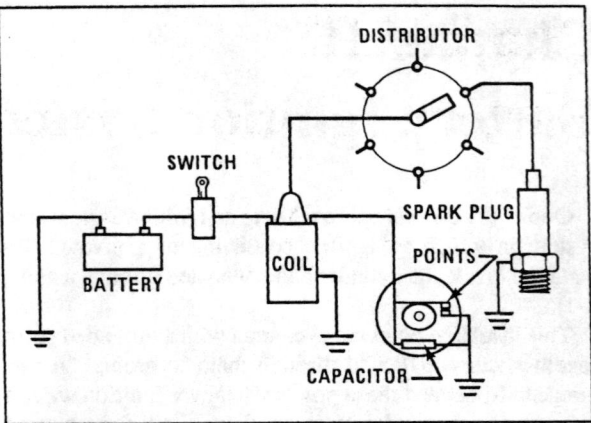

**Figure 2-2.
Battery ignition
system.**

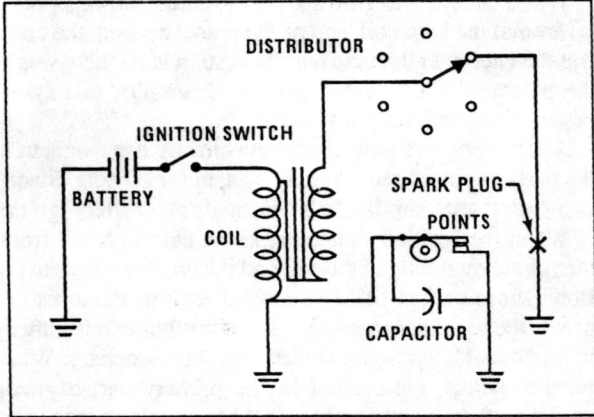

**Figure 2-3.
Schematic of a
battery ignition
system.**

turns of the secondary induces a high voltage in it. The distributor rotor is so timed that when the points break, it is aligned with the lead to the proper spark plug, and the high voltage will cause an arc across the spark plug electrodes, igniting the fuel-air mixture in the engine cylinder.

Primary current flows during the time the breaker points are opening. If provisions are not made to quench it, an arc will be drawn across the points, and they will soon burn and weld themselves together. A capacitor is connected in parallel with the points, so that as they begin to open and the resistance starts to build up, electrons follow what appears to be an easier path to ground, through the capacitor, and current immediately stops flowing at the points.

By the time the capacitor is fully charged, the points are opened wide enough that no current can flow, and so the primary field collapses almost instantly.

QUESTIONS:

1. *Does the spark from a battery ignition system become stronger or weaker as the engine speed increases?*

4

Chapter III
Magneto Ignition Systems

Battery ignition systems have an advantage over magneto systems for starting, because at slow speeds the primary current has plenty of time to build up to maximum and create a hot spark for starting the engine. But there is a corresponding disadvantage: as the engine speed increases, there is insufficient time for a complete buildup of the primary magnetic field, so the intensity of the spark decreases with engine speed.

Magnetos have been developed to provide a simple, dependable ignition system, completely independent of the electrical system of the airplane. They differ from the battery system in that the primary current is interrupted by a set of breaker points, and high voltage for the spark plugs comes from the voltage step-up in the magneto coil.

A. Basic Forms Of Magnetos

1. Shuttle-type Magneto

There have been three types of AC generators used in aircraft magnetos, but, as with many other aspects of aviation, these have been reduced to a single form. Some of the very earliest magnetos used a fixed magnet and a rotating coil, similar to the magneto used in the early telephones. The primary and secondary coils as well as the capacitor (formerly called a condenser) all rotated, and a set of contacts opened at the proper time to interrupt the flow of primary current and induce a high voltage into the secondary winding.

The fact that the coils and capacitor were part of the rotating element brought problems, and so magnetos with fixed coils were developed.

2. Polar Inductor Magneto

Polar inductor magnetos, such as those made by the American Bosch Co., use fixed permanent magnets and fixed coils. But they obtain the reversal of flux by use of a soft iron rotor, which alternates the route the flux takes through the coil core (Fig. 3-1).

Figure 3-1A: A soft iron rotor completes the magneto circuit so lines of flux can pass through the coil core from right to left.

Figure 3-1B: Ninety degrees of magneto rotation later, the soft iron rotor completes the magnetic circuit again, but this time the flux lines pass through the coil core from left to right.

Figure 3-1.
Polar Inductor
magneto.

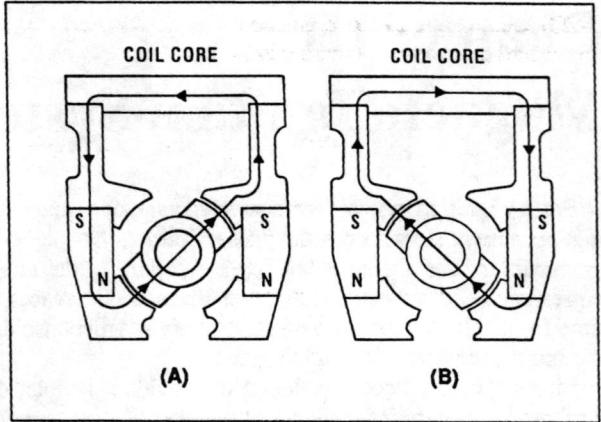

(A)   (B)

3. Rotating Magnet Magneto

By far the most popular magnetos in use today have fixed coils and capacitors and rotating permanent magnets. This configuration was pioneered by the Scintilla Corporation, now the Electrical Components Division of the Bendix Corporation.

B. Magnetic Circuit

Figure 3-2 shows the magnetic circuit of a typical four- or six-cylinder, high-tension magneto.

In Fig. 3-2A, the two-pole magnet is in its full register position with the north pole on the left side. Maximum flux now flows through the soft iron frame and coil core in a clockwise direction. As the magnet rotates, the amount of flux decreases until at the neutral position (Fig. 3-2B). There is no flux in the frame or core, but continued rotation brings the magnet again to full register (Fig.

Figure 3-2.
Magnetic circuit of a
rotating magnet
magneto.

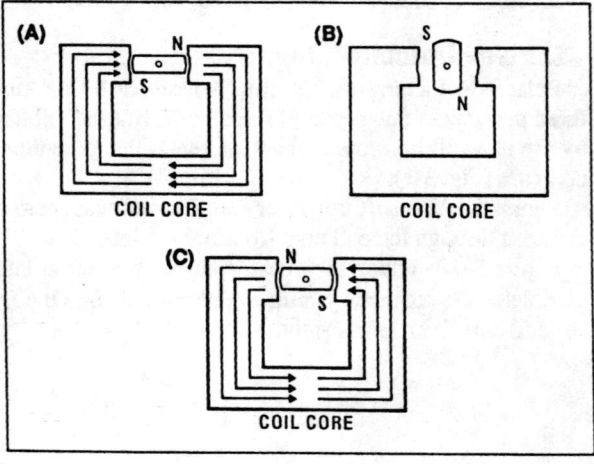

3-2C), this time with the opposite polarity. The flux is again maximum in the frame and coil core, but is now in the opposite direction.

1. Static Flux

A plot of the static flux relative to the magnet position is shown in Fig. 3-3A. Flux is maximum at full register, minimum at neutral, and again maximum, but in the opposite direction, at full register.

Fig. 3-3A. Static flux: The rotating magnet would produce a smooth curve similar to this if there were no windings around the coil core.

Fig. 3-3B. Primary current: Caused to flow by the changing flux in the coil core. The current stops abruptly when the points open.

Fig. 3-3C. Resultant flux: The flux caused by the primary current delays the change in the flux from the magnet until the points open. The E-gap angle is the number of degrees the magnet rotates beyond the neutral position before the breaker points open.

2. Primary Current

Current generated in an electromagnetic generator, of which the magneto is a form, is determined by the *rate* at which the lines of flux change or cut across the windings of the coil. The primary coil is wound over the soft iron core, and in the full-register position of the magnet, the flux is maximum. But the *change* is minimum, so no current is being generated in the primary winding. As the magnet leaves the full-register position, there is no flux, but since it immediately starts to build in the opposite direction, the *change* in flux is maximum. And it is at the neutral position that the maximum amount of primary current *should* be caused to flow.

3. Resultant Flux

The condition which at first glance should exist actually does not, and for a good reason. As the flux begins to change, current starts to flow in the primary

Figure 3-3.
Flux curves in an
aircraft magneto.

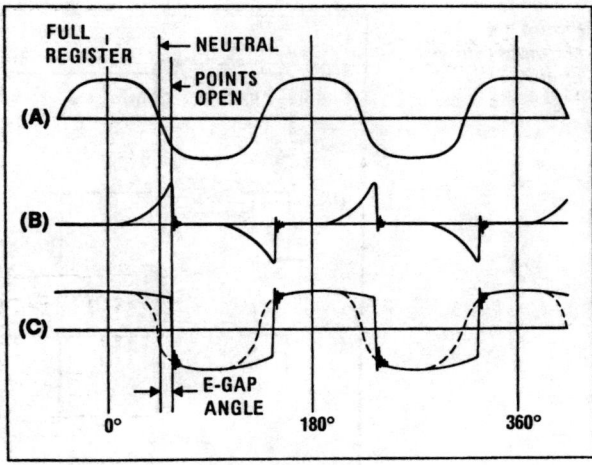

7

winding, and this current also causes a magnetic flux which, according to Lenz's law, *opposes* the flux which caused the current. The result of the primary flux reacting with the static flux is seen in the curve, Fig. 3-3C, the resultant flux. The result is a delay in the buildup of the primary current. So instead of the maximum current flowing at the neutral position of the magnet, it flows several degrees after neutral.

The desired end result of a magneto is to have the maximum amount of voltage generated in the secondary winding of the coil. This voltage is determined by the rate of collapse of the magnetic field of the primary. In order to achieve the maximum rate of collapse, the primary current must be stopped *instantaneously* at the point of its maximum flow.

C. Electrical Circuit

1. Primary Circuit

In Fig. 3-4 we see a portion of the primary circuit of a magneto.

Fig. 3-4A: When the points are closed, the current generated by the rotating magnet flows to ground.

Fig. 3-4B: As the points begin to open, current flows into the capacitor which appears to be a path to ground.

Fig. 3-4C: By the time the capacitor has charged, the points have opened far enough that current cannot flow through them, and no arcing will occur.

Current is generated in the primary coil and flows through the breaker points to ground, until the cam begins to open the points. The electrical inertia of the primary current tries to keep it flowing as the resistance across the points increases. Remember the effect of inductance?

When the current starts to decrease, the magnetic field around the coil begins to collapse. The collapse, or change, generates a current which *opposes the decrease* or attempts to keep the current flowing. If you will look back to Fig.

Figure 3-4.
Magneto primary circuit.

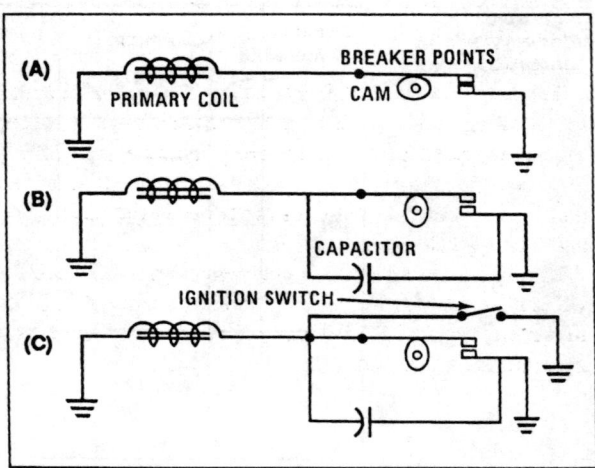

8

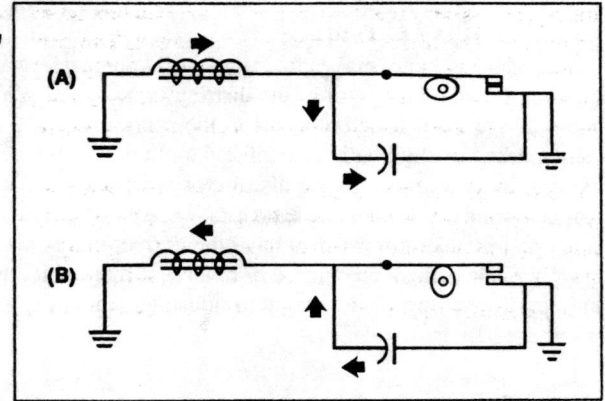

Figure 3-5.
The capacitor aids in the buildup of the primary current.

(A)

(B)

2-1 you will see that this circuit is similar to the kick coil in the make-and-break igniter of the early system. We wanted an arc then, but in this magneto, an arc is exactly what we *do not* want.

To prevent an arc at the points, a condenser, more accurately called a capacitor, is placed in parallel with them (Fig. 3-4B). As resistance across the points increases, electrons flow into the capacitor, since it *appears* to be a lower resistance path to ground. By the time the capacitor is charged, the points have opened far enough that electrons cannot flow across. The primary current has effectively been stopped instantaneously.

The capacitor not only prevents arcing of the points, but helps get a good flow of primary current started. The magneto produces alternating current, so consecutive sparks have alternate polarity.

In Fig. 3-5A, let's assume that electrons have flowed into the capacitor as shown by the arrow. The capacitor charges up and then seeks to discharge through the points. But, as we have seen, the points are open by this time, and there is no path to ground through them. But the next half-cycle of alternating current has started to build up (Fig. 3-5B). The capacitor can now discharge through the primary coil and, in so doing, help get the flow of primary current going in the right direction.

The magneto coil consists of a laminated soft iron core that is tightly wedged into the magneto frame. Around this core is wound a primary coil, consisting of several hundred turns of relatively heavy wire. On top of this are several thousand turns of very small wire. The voltage buildup in this coil is dependent on both the rate of collapse of the flux and the turns ratio between the primary and the secondary winding.

The ignition switch for a magneto, Fig. 3-4C, is in the primary circuit and when it is open, the magneto is "hot" or operational. When the switch is placed in the "off" position, the primary circuit is grounded, and no voltage can be built up in the secondary coil.

2. Secondary Circuit

Secondary voltage is taken from the high-voltage terminal of the coil through a carbon brush, to the rotor of the distributor, and then across an air gap to the high-voltage lead, which attaches to the spark plug. The distributor gear has teeth marked to align with a chamfered tooth on the magneto drive gear. There are usually two marks on the distributor gear; one for the right-hand rotation magnetos and one for the left-hand rotation. And when assembling the magneto, care must be taken in order to have the correct marks aligned. Provisions are usually made using a chamfered or otherwise marked tooth, visible through an inspection opening in the magneto housing, to identify the number one distributor position.

D. Special Forms Of Magnetos

1. Dual Magneto

In the interest of both safety and better combustion characteristics in the cylinder, all certificated aircraft engines have dual ignition. This requires two separate and independent ignition systems. Some early engines used a combination of battery ignition and a high-tension magneto; the battery system for starting, and then both systems for normal operation. Two separate magnetos are used on most modern engines, but with the need for more accessories and the limited number of drive pads available, the dual magneto is becoming popular.

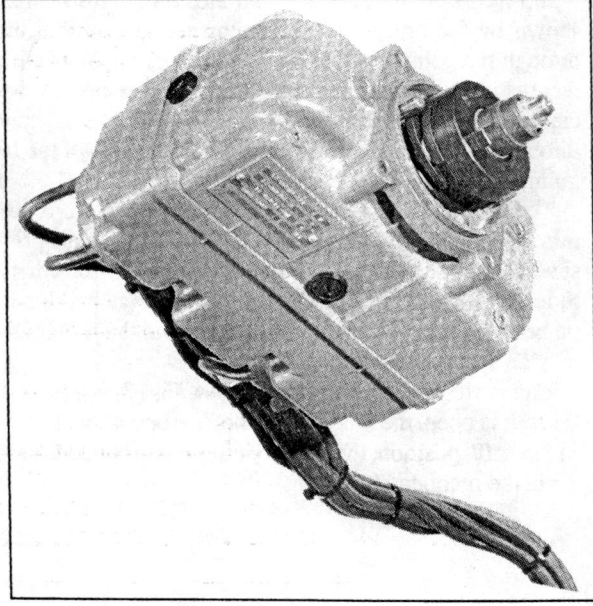

Figure 3-6.
The Bendix D-3000 dual magneto is actually two separate ignition systems, having only the rotating magnet, cams, and housing in common.

This concept is not new, as dual magnetos were used on radial engines in World War II, including the Pratt and Whitney R-4360, a twenty-eight cylinder, four-row radial engine, which used seven dual, four-cylinder magnetos mounted around its nose section.

Dual magnetos may be considered as two separate ignition systems, as only the housing, rotating magnet, and cam are common to both systems. There are two sets of breaker points, two coils, two capacitors, and two distributors. The magnetos are similar in principle and operation to single magnetos. Only in their timing to the engine are they unique. Their timing will be covered in detail in the appropriate section of this book.

2. Low-tension Magneto

The majority of the magnetos used on general aviation engines are of the rotating-magnet, high-tension, single-magneto configuration, but there are others.

One of the problems inherent with high-tension magnetos has been the inability of the harness and distributor to contain high voltage in the thin air of high altitudes. As the air becomes less dense, its insulating capability decreases, and sparks jump from the distributor finger to the housing, or an electrical

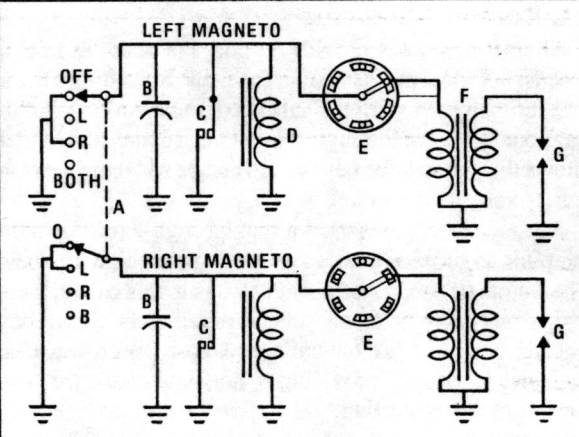

Figure 3-7.
Low-tension ignition system.

A- IGNITION SWITCH: ONE SWITCH FOR THE
   COMPLETE SYSTEM.
B- CAPACITORS: ONE IN EACH MAGNETO.
C- BREAKER POINTS: ONE SET IN EACH MAGNETO.
D- PRIMARY COILS: ONE IN EACH MAGNETO.
E- CARBON BRUSH DISTRIBUTOR: ONE IN EACH
   MAGNETO.
F- HIGH-TENSION TRANSFORMER: ONE FOR EACH
   SPARK PLUG. THESE ARE USUALLY LOCATED ON
   THE CYLINDER HEAD, NEAR THE SPARK PLUG.
G- SPARK PLUG: TWO IN EACH CYLINDER.

breakdown occurs within the harness itself. Steps have been taken to prevent this, such as making the distributor physically larger, so there will be a greater distance for the spark to travel. Magnetos, distributors, and harnesses have been pressurized with compressed air to increase their resistance to breakdown. But one of the more practical developments, which has allowed magneto ignition to function at high altitude, has been the low-tension magneto.

Similar to the high-tension magneto in the generation of primary current, the low-tension magneto has a rotating magnet, cam-operated breaker points, and a capacitor. But it differs in the production of the spark in the secondary.

The coil in the low-tension magneto has only the primary winding, with its output going to a carbon-brush distributor. Primary current is carried to individual coils or transformers, one for each spark plug, mounted on the cylinder near the plug. When the breaker points open, the primary current through a specific coil is interrupted, its field collapses and generates a high voltage in the secondary winding, which is taken to the spark plug by a very short lead.

E. Aids To Starting

1. Booster Magneto

Aircraft magnetos provide a good, hot spark at idle, at cruise, and at high speeds; but some provision must be made for getting a hot spark when the engine first turns over in starting. This spark must not only be hot, independent of the rotational speed of the engine, but it must come later than the normal spark. This allows the piston to be beyond top center when the pressure from the expanding gases exerts its push on it.

Some early engines used a regular high-tension magneto, hand-cranked by the pilot, to provide a hot spark while the engine was being cranked (Fig. 3-8). The output of this booster magneto, as it was called, went directly to a trailing finger on the rotor of one of the distributors. This allowed the high-voltage booster output to fire the cylinder whose piston was near top center or which had just passed it, providing a hot, late spark *for* starting, and preventing kick-back *while* starting.

Figure 3-8.
Hand-cranked
booster magneto.

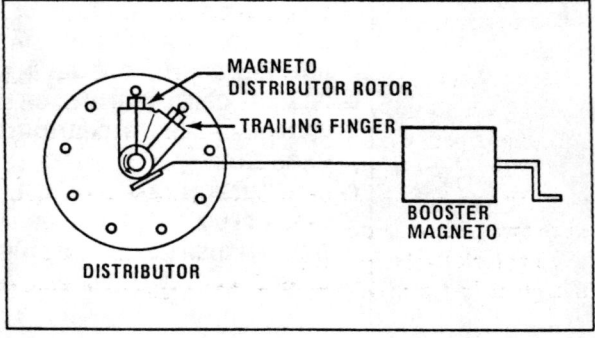

Figure 3-9.
Impulse coupling.

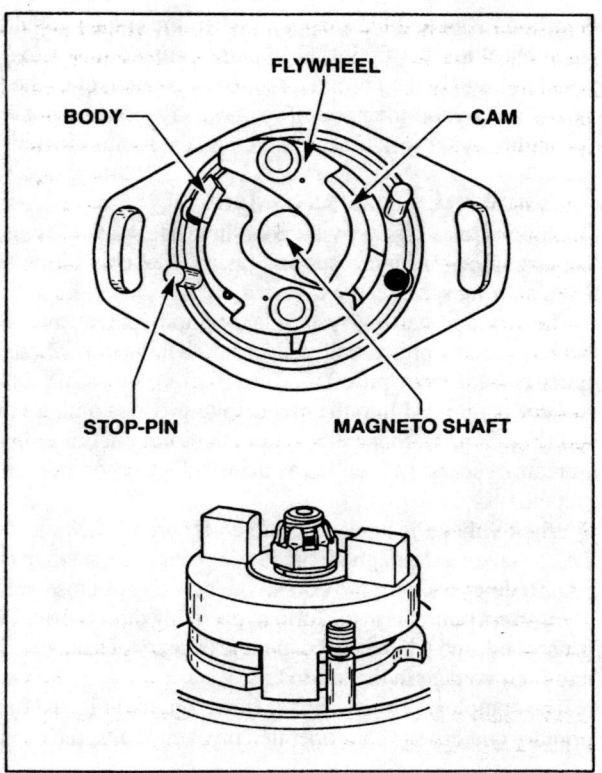

2. Impulse Coupling

As smaller engines became popular in general aviation, starting procedures were simplified, and the booster magneto was replaced with the impulse coupling (Fig. 3-9).

A cam plate with two flyweights attached is keyed to the magnet shaft. The impulse coupling body rides over the cam and flyweight assembly and is the part driven by the engine.

The body and cam plate are connected through a heavy-duty, clock-type spring. Two stop pins are placed in the magneto housing, in such a position that as the starter rotates the engine, the flyweights move out and contact the stop pins.

This holds the magnet and cam plate still as the engine continues to turn, winding the spring. After a predetermined amount of rotation, projections on the coupling body contact the flyweights and release them from the stop pins. The spring now spins the magnet, producing a hot, late spark. As soon as the engine starts to fire and pick up speed, centrifugal force on the flyweights holds them away from the stop pins.

They lock the body and cam plate together so that the magneto operates as though it had a solid coupling. Magnetos with impulse couplings may be identified by an audible snapping, when the engine is pulled through by hand. When checking the timing of impulse coupled magnetos, the engine must be pulled through in the normal direction of rotation until the coupling snaps, then backed up beyond the point of ignition advance, but not enough to engage the flyweights. After this, timing is done in the normal way.

3. Shower Of Sparks System

Impulse couplings provide good hot, late sparks for starting small engines. But as engine size increases and operational conditions become more severe, better starting systems are required.

The vibrator starting system is an improvement over the booster magneto, and has become pretty well standard. The induction vibrator produces pulsating direct current from pure DC (Fig. 3-10A). Pulsating DC from the induction vibrator is directed into the magneto to produce a high voltage in the magneto coil (Fig. 3-10B). Since this system does not put out a single spark, but instead a stream of sparks, it is called by the Bendix Corporation the "Shower of Sparks" system.

When voltage is applied to the coil, current flows and energizes it, making the core an electromagnet. The lower point is attracted to the core and opens the circuit, de-energizing the coil so the points will close and start the cycle over. The current output is in the form of pulsating direct current with about 200 pulses per second, and when it is fed into the primary of a magneto coil it is transformed into high voltage in the secondary.

The complete primary circuit (refer again to Fig. 3-10) includes not only the primary coil but the cam-operated breaker points, the capacitor and the ignition

Figure 3-10.
Induction vibrator
system.

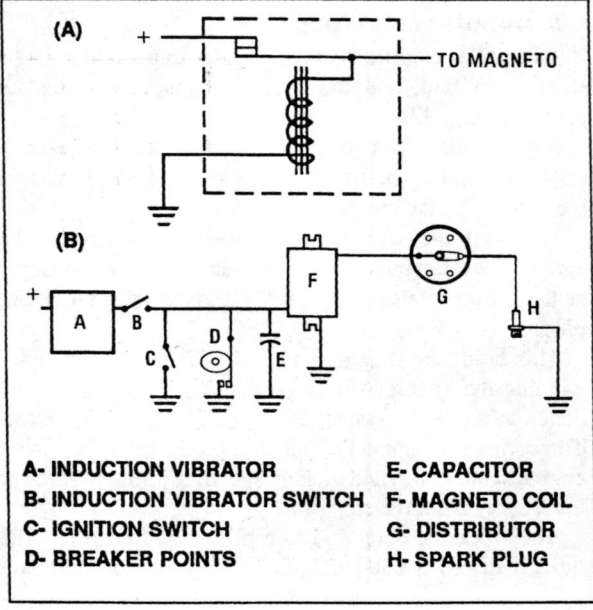

A- INDUCTION VIBRATOR   E- CAPACITOR
B- INDUCTION VIBRATOR SWITCH   F- MAGNETO COIL
C- IGNITION SWITCH   G- DISTRIBUTOR
D- BREAKER POINTS   H- SPARK PLUG

14

switch. With the ignition switch closed there is still the rise and fall of primary current, but there is not the sudden collapse needed to induce a secondary voltage.

For an aircraft engine to start, there must be a hot and *late* spark, and since aircraft magnetos have fixed timing, the normal spark cannot be retarded. To prevent the engine attempting to kick back from a spark occurring at its normal advance position, the Shower of Sparks must be timed so it will not fire the spark plug until the piston is at or near top dead center (Fig. 3-11).

If the output of the vibrator is fed into the primary of the magneto coil, voltage will be induced into the secondary, but when the breaker points short the pulsating DC to ground, no voltage will be induced in the secondary. As soon as the points open, though, the current will go to ground through the primary and sparks will be produced as long as the points remain open. This produces the required hot spark, but does not help in getting the late spark. To accomplish this, a second set of points, the retard points, are installed in one of the magnetos, usually the left one. When the engine is being started, pulsating direct current goes to ground through both the run and the retard points which are in parallel.

The run points open first, at the normal advance position, but since the retard points are still closed, there is no spark at the secondary. At the proper position for the starting spark to occur, the retard points open, and the only path for the pulsating direct current is through the primary of the coil to ground. This provides a continual spark until the run points close. During this time, the distributor rotor is aligned with the electrode for one of the ignition leads.

Figure 3-11.
Provisions for obtaining a late spark with an induction vibrator system.

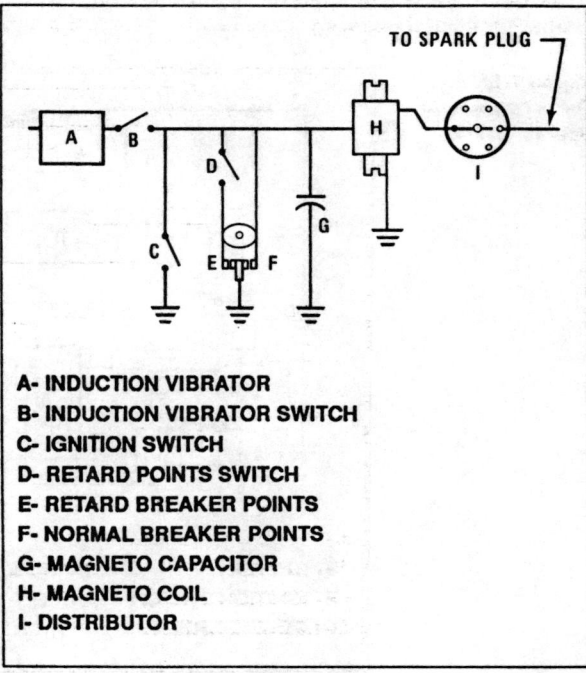

A- INDUCTION VIBRATOR
B- INDUCTION VIBRATOR SWITCH
C- IGNITION SWITCH
D- RETARD POINTS SWITCH
E- RETARD BREAKER POINTS
F- NORMAL BREAKER POINTS
G- MAGNETO CAPACITOR
H- MAGNETO COIL
I- DISTRIBUTOR

15

The basic function of the Shower of Sparks system is rather simple (Fig. 3-12). But there are other things involved in starting an aircraft engine, besides providing the spark and to simplify starting. Some of these functions are incorporated in the ignition switch circuit, so when the switch is placed in the "Start" position, the following things happen:

1. The right magneto is grounded to prevent a spark at the advanced position, which could cause a kick-back and possibly damage the starter.
2. The retard points circuit is completed so they become operational.
3. The vibrator is energized, sending pulsating direct current into the magneto.
4. The starter solenoid is energized, cranking the engine.

The ignition switch for a Shower of Sparks system is naturally more complicated than one for just the magnetos alone. Fig. 3-12 shows the entire circuit. The switch is shown as a series of simple switches. But a rotary type switch is normally used, one with an Off position fully counterclockwise, then Right, Left, Both, and a spring-loaded Start position.

Fig. 3-13 shows the switch conditions for the various positions.

A. In the Start position, the starter solenoid and vibrator are energized, and the retard points are connected to the vibrator. The right magneto is grounded, and the left is operational.
B. In the Both position, all of the switches are open. This allows both magnetos to be operational, with the retard points out of the circuit of the left magneto.
C. In the Left position, all of the switches are open except the one which grounds the right magneto.
D. In the Right position, all of the switches are open except the one which grounds the left magneto.

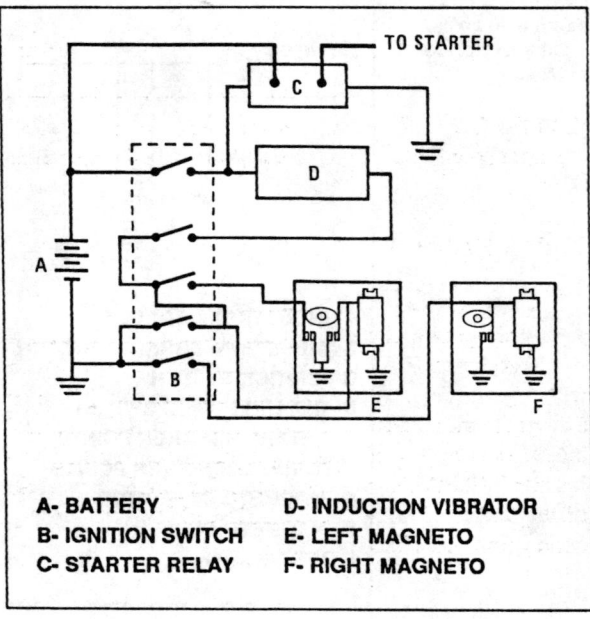

Figure 3-12.
Basic "Shower of
Sparks" system.

TO STARTER

A- BATTERY   D- INDUCTION VIBRATOR
B- IGNITION SWITCH E- LEFT MAGNETO
C- STARTER RELAY   F- RIGHT MAGNETO

Figure 3-13.
Switch positions for
Shower of Sparks
system: (A) Start;
(B) Both; (C) Left;
(D) Right; (E) Off.

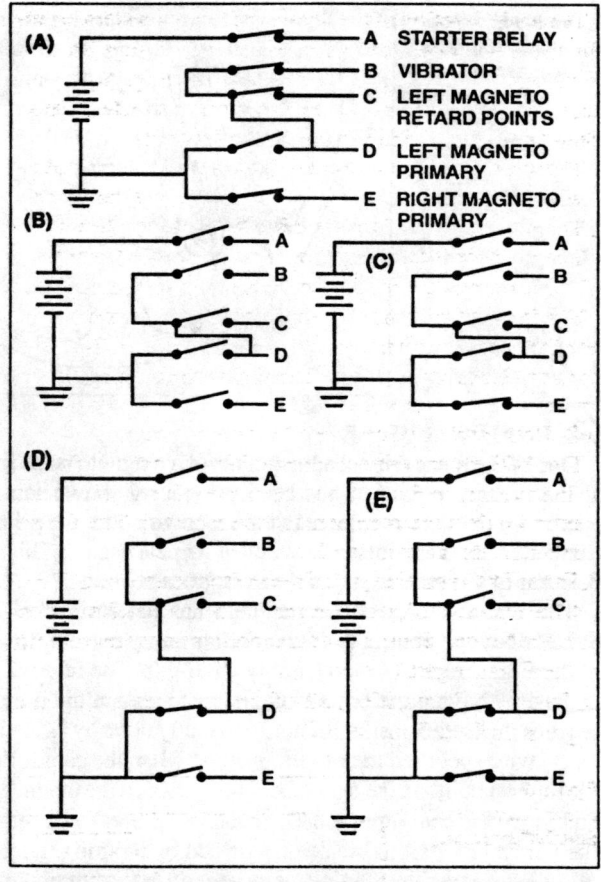

A STARTER RELAY
B VIBRATOR
C LEFT MAGNETO
  RETARD POINTS
D LEFT MAGNETO
  PRIMARY
E RIGHT MAGNETO
  PRIMARY

E. In the Off position, both magnetos are grounded, and the switches to the starter solenoid and vibrator and those in the retard circuit are open.

F. Magneto Servicing

1. Preparation For Installation

The simplicity of the magneto is one of the reasons for its popularity as a spark producer. While the magneto has fewer parts, those which it has are critical with regard to timing and are quite intolerant of poor service. Before installing a magneto on an aircraft engine, be sure that the particular part number of the magneto is approved and appropriate for the engine on which it is being installed. Check the entire magneto for physical condition. There should be no oil or grease inside the case where it could foul the points, and all of the electrical connections should be secure.

17

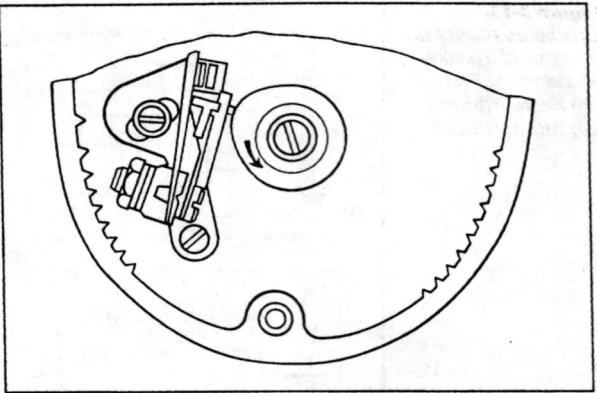

Figure 3-14.
The manufacturer's
service manual will
specify the maximum
amount the points
should open.

2. Internal Timing

One of the more critical adjustments of a magneto is the timing of the opening of the breaker points. It has been previously shown that in order to get the maximum buildup of voltage in the secondary coil, the primary current must be stopped at the very instant it reaches its peak value. This is not at the neutral position of the magnet, but several degrees beyond.

The manufacturer of the magneto has determined exactly the number of degrees beyond neutral the maximum primary current flows, and this is known as the E-gap angle, or more simply the E-gap. The internal timing of a magneto is done by having the breaker points just break at the time the rotating magnet reaches the E-gap position. This is normally done by adjusting the points so they will have a specified maximum opening when the cam follower, Fig. 3-14, is at the highest point on the cam lobe. Then position the magnet in its E-gap position, and move the cam on the shaft, or rotate the breaker assembly on its base, until the points just start to break as indicated by a timing light.

After making this adjustment exactly as recommended by the magneto manufacturer, turn the magneto away from E-gap, and as you rotate it back, the timing light should indicate that the points are just breaking as E-gap is reached. When the points are opened their maximum, the clearance should be within the limits set by the manufacturer. But remember, it is when the points open that is critical, not how much they open (Fig. 3-15).

3. Magneto To Engine Timing

In addition to timing the magneto internally, it must provide its hot spark when the piston is in its correct position in the cylinder.

Maximum cylinder pressure should be reached shortly after the piston passes top center (Fig. 3-16). If the spark occurs too soon, maximum pressure will be built up before the piston reaches top center. In starting, the engine will kick back, or if it is running, there will be a serious loss of power. If the spark occurs too late, the piston will already be on its down stroke, and power will be lost. The mixture will still be burning when the exhaust valve opens, and overheating

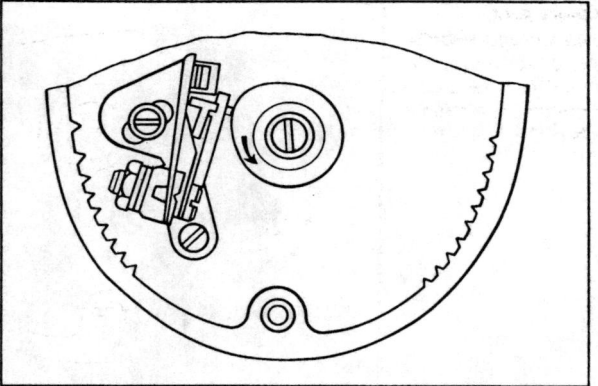

Figure 3-15.
Critical to magneto timing is the position of the rotating magneto at the instant the points break.

and valve burning will result. Normal ignition occurs somewhere around twenty to thirty degrees before the piston reaches top center on the compression stroke.

To properly time the magneto to the engine, the engine is positioned with number one cylinder at its firing position, and the magneto turned to its E-gap position with the distributor finger in position to fire number one cylinder. The distributor gears are marked with a chamfered or otherwise identified tooth, visible through an inspection window when the distributor is in number one firing position.

With the magneto in its proper firing position, and the engine properly positioned, the magneto is installed and secured. Rotate the engine a couple of revolutions, until the number one cylinder comes up in firing position again. Check with a timing light to be sure that the points just break at the proper number of degrees of crankshaft rotation, with the marked tooth in the inspection window (Fig. 3-17).

Some of the larger radial engines have compensated cams, cams with one lobe for each cylinder, to compensate for the fact that only the master rod rotates

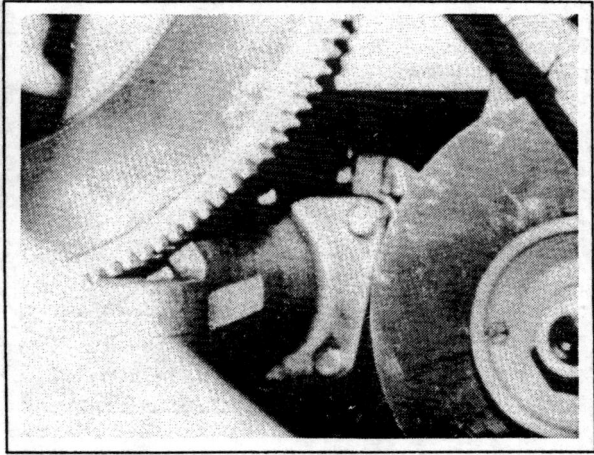

Figure 3-16.
The crankshaft of the engine should be placed so that number one cylinder is on the compression stroke, the proper number of degrees before top center.

19

Figure 3-17.
The marked tooth on the distributor gear identifies the position of the rotor for firing cylinder number one.

around the crank shaft, and all of the other pistons attach to it and rock back and forth or articulate (Fig. 3-18). The cam is ground in such a way that the breaker points open, not at the same angular position of the crankshaft, but at the same linear distance of the pistons from the top of their stroke.

The master-link rod arrangement causes the angular and linear distances to be different for each cylinder. The lobe for cylinder number one is marked. Either a step or a slot is cut across the cam so a straight edge can be aligned with a mark in the magneto case for timing.

Magnetos equipped with impulse couplings present little special problem when timing them to the engine as long as a simple procedure is followed. After the magneto is installed in its approximate firing position, rotate the engine until the impulse coupling snaps and then back it up until the breaker points close. Now, turn the engine forward until the points just open. Caution should be observed when timing a magneto with an impulse coupling that the timing light

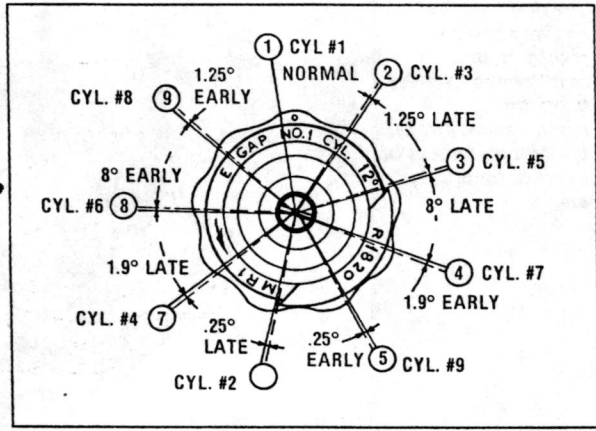

Figure 3-18.
High-performance radial engines use a compensated cam. The master-link rod arrangement causes the angular and linear distances to be different for each cylinder.

20

is not connected to the magneto when the impulse coupling snaps. The magneto is spun so fast that the primary voltage can damage some types of timing lights.

The Slick magneto, manufactured by Slick Electro, Inc., is a popular magneto found on many of the current general aviation aircraft. While the principles involved in its operation are the same as those of the Bendix magneto, the methods of its timing differ enough that it is important that we understand them.

The rotating magnet assembly and the frame are drilled in such a way that when the magnet is in its E-gap position you are able to insert a special timing pin or a sixpenny nail through the frame into the magnet assembly. This holds the magnet while the points are adjusted so that they just begin to open.

Align either the LH or RH mark on the distributor gear with the marked tooth on the drive gear to time the distributor to the magneto, and install the distributor housing, which is the back half of the magneto. Nylon is used for both the drive and distributor gears. Because of nylon's dimensional change with temperature, the mesh of these gears is extremely loose when the magneto is cold, but as the magneto gets up to operating temperature, the mesh becomes normal.

When timing a Slick magneto to the engine, the vent plug on the side of the housing is removed, and the magneto is rotated until the marked tooth on the distributor gear for the proper direction of rotation is aligned with the mark in the housing. In this position, a timing pin or sixpenny nail can be slipped through the housing into the rotating magnet assembly and will hold the magneto in its E-gap position (Fig. 3-19).

With the engine in position for number one cylinder to fire, the magneto can be slipped into place, secured, and the timing pin removed. Rotate the engine and bring it back into firing position for number one cylinder, and a timing light across the breaker points will indicate the exact position the points open. If the points do not open at the exact firing position, the magneto can be slightly loosened and tapped enough to cause them to just begin to open.

One of the unique innovations of Slick Electro, Inc. is their nonserviceable magnetos. Their 4000 series lightweight magnetos are sealed and no replace-

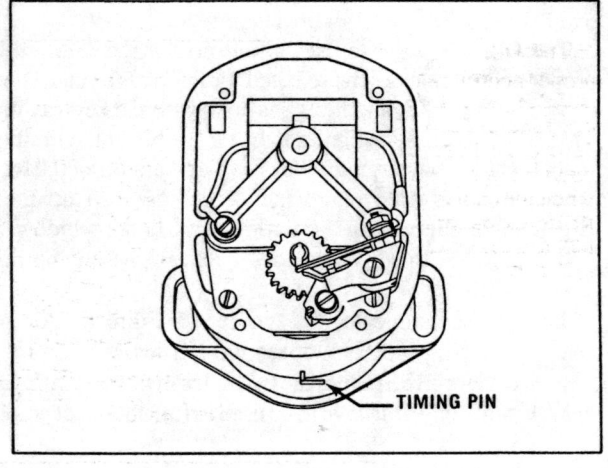

Figure 3-19.
Timing pin inserted into the rotating magnet of a Slick-Electro magneto.

TIMING PIN

ment parts are available. They are engineered so the wear of the breaker cam exactly balances the wear of the points and the timing will stay the same for the entire engine overhaul period. At the time of engine overhaul, the magnetos are returned to the factory and exchanged for factory rebuilt units.

When installing this series magneto, they should be "sparked out" to determine the proper magnet position for installation. Hold the number one lead about a sixteenth of an inch from the magneto frame and rotate the drive until the impulse coupling snaps, and a good healthy spark jumps from the number one lead to the case. This indicates that the magneto has just fired number one plug, so the magnet is backed up enough to align the hole in the rotating magnet with the hole in the frame. Insert the timing pin or a sixpenny nail and install the magneto on the engine after the piston of number one cylinder has been put in the firing position.

4. Dual Magneto Timing And Installation

Dual magnetos such as the Bendix D-2000 offer little in the way of special problems, as it is actually two separate magnetos, sharing only the case, rotating magnet, and cams. There are enough special features, however, with regard to timing this magneto that it is included here.

For uniformity, the top of the magneto is the side with the name plate, and the lower set of points, the left points, are used as the reference or datum points when timing this magneto.

By removing the rear cover, the cams, breaker assemblies and distributor blocks are accessible. In addition to the two main breaker assemblies, the retard breakers for the Shower of Sparks starting system may be stacked on an extended bracket above the left main breakers, and a second cam stacked on the shaft to operate them. In certain instances, even another set of points may be stacked on an extension bracket over the right main breakers to operate an electronic tachometer.

The capacitors used in these magnetos serve a dual purpose. They are in series with the primary switch lead to filter out radio interference, as well as being in parallel with the points for arc suppression.

The D-2000 magneto uses a four pole magnet, and it is important that the proper neutral position be selected for internal timing. The magnet is marked as shown in Fig. 3-20, and timing is done with the keyway in the magnet shaft *up*. This is indicated by the letter K being visible in the timing window. Adjust the main breaker contacts until they have a clearance of 0.016, plus or minus 0.002, when the cam is at the highest part of the lobe; then turn the magnet so the proper E-gap angle aligns with the timing mark in the window, and lock the shaft in this position with a friction lock between the engine drive member and the housing.

Loosen the main cam, and turn it in the direction of magneto rotation until the approaching lobe just opens the left main breaker. (The tips of a pair of Tru-Arc pliers inserted into the two holes in the cam make a good tool for turning it.) A timing light will have to be used to find the exact point at which the contacts

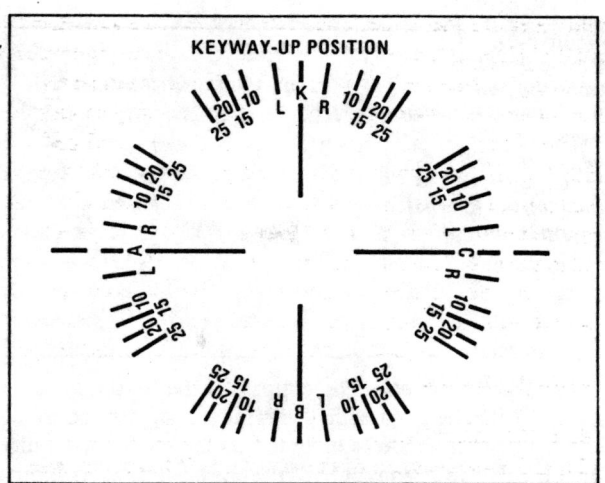

Figure 3-20.
Magnet of the Bendix
D-2000 magneto.

break. Adjust the right main breaker to open at the same time by slightly loosening the breaker hold-down screws, and, using a small drift, gently tap the breaker frame toward the cam until the timing light indicates these points have just opened. Torque the cam and breaker screws, loosen the friction lock, and recheck the E-gap and breaker synchronization. Readjust if necessary by repeating the breaker opening procedure.

Adjust the retard points so they will open a maximum of 0.016", plus or minus 0.004", with their follower at the highest point of the cam lobe. Now turn the magneto shaft to the proper position for the retard points to open, and lock it in place with the friction lock. Loosen the cam screw just enough that the top cam can be moved, and turn it in the direction of magneto rotation until the timing light indicates that the retard points are just opening. Torque the cam screw as indicated in the service manual, loosen the friction lock, and check the points opening. It should be within the tolerance allowed by the manufacturer.

Be sure that the capacitor wire is positioned and formed so it will not interfere with the high-tension wells or the cover screws, and the high-tension lead grommets are all seated. The cover can then be installed and the cover screws all torqued according to the recommendations in the service manual.

The magneto is installed on the engine by placing number one piston in the proper firing position and the magneto in the proper E-gap position, as indicated by the marked teeth visible in the windows at each end of the magneto housing. These marked teeth will appear every time the distributor is in position to fire number one cylinder; and to be sure that the neutral selected is the one with the keyway up, the letter K should be visible in the inspection window at the top of the magneto. If the top of the magneto is inaccessible, the inspection plug on the bottom of the magneto may be removed, and the letter B should be visible opposite the timing mark in this window.

When the magneto cover is in place, a jumper will have to be used to connect the timing light into the P lead circuit, and when installing the jumper, be sure

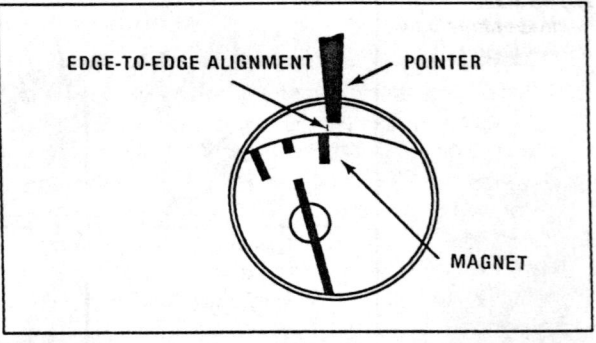

Figure 3-21.
If the points break while the pointer on the housing is within edge-to-edge alignment with the proper mark on the magnet, the internal timing is correct.

EDGE-TO-EDGE ALIGNMENT

POINTER

MAGNET

to use the correct one. The jumper for the S-20 magneto will fit into the hole, but since it is longer, it will distort the spring at the bottom of the terminal. When the magneto is securely mounted on the engine it should be so timed that the first breaker to open will be at the engine timing position; and if the two sets of points do not open at exactly the same time, the later points should open within three engine degrees of the first. Failure for both sets of points to open within this tolerance could indicate timing problems inside the magneto, and it should be removed from the engine and the trouble corrected.

It is possible for magneto timing to drift either early or late, due to cam follower or point wear, and the magneto may be loosened and very slightly bumped, to cause the points to open at the correct time relative to the piston position. This dual magneto may be bumped a maximum of 3/32 inch for the four-cylinder models and 1/16 inch for the six-cylinder models. Another indication of the amount the magneto has been bumped is the relationship between the E-gap angle marks and the timing pointer (Fig. 3-21). If you have not exceeded the edge-to-edge distance between the pointer and the E-gap mark on the magnet, the magneto has not been bumped too much.

QUESTIONS:

1. What provision is made in a magneto or battery ignition system to minimize the arcing across the breaker points?

2. Does the spark from a magneto become stronger or weaker as the engine speed increases?

3. Is the primary current in a magneto pulsating direct current or alternating current?

4. At which position of a rotating magnet is the change type the greatest?

5. Why does the maximum primary current not occur at the neutral position where the maximum rate of static flux occurs?

6. Is the primary capacitor in series or in parallel with the breaker points?

7. Of what material is the core of a magneto coil made?

8. When a magneto switch is open, is the magneto "on" or "off?"

9. How does a dual magneto differ from a single magneto?

10. How can a low-tension magneto produce a spark in an engine cylinder?

11. Why is there little danger of flash-over in the distributor of a low-tension magneto?

12. Where does a booster magneto direct its output?

13. What is the purpose of an impulse coupling?

14. What is the purpose of the extra set of breaker points in a magneto equipped for the Shower of Sparks starting system?

15. Name four things that happen when the ignition switch for an engine equipped with the Shower of Sparks system is put in the "Start" position.

16. What is the condition of the primary current at the moment the breaker points open on a properly timed magneto?

17. Which is more important in checking the timing of a magneto, the time the points open, or the amount they open?

18. Where is the piston in the cylinder when normal ignition occurs?

19. Who do high-performance radial engines use compensated cams in their magnetos?

20. What should be done to an impulse coupling on a magneto when timing the magneto to the engine?

21. How is a Slick magneto held in its E-gap position when the magneto is installed on the engine?

22. What provision is made in the nonserviceable-type Slick magnetos to maintain the proper internal timing for the entire engine overhaul period?

23. What are the two functions of the capacitor in a Bendix D-2000 dual magneto?

24. Are the retard points opened by the same cam as the regular points on a Bendix D-2000 dual magneto?

Chapter IV
Ignition Leads

Enough high voltage may be generated in a magneto to provide sufficient spark for igniting the fuel-air mixture, but if it is not carried to the spark plug without losses, the engine performance will deteriorate.

Modern aircraft carry a considerable amount of electronic equipment, and the communications and navigation systems must be able to receive signals from ground stations without interference from extraneous sources. Since high-voltage ignition systems constitute very effective radio transmitters, the energy radiated from the spark must be contained within the harness and grounded. Unless this is done, there will be enough interference to impair radio reception.

A. Construction

Ignition leads are usually made of stranded copper or stainless steel wire with rubber or silicone insulation, and almost universally now, there is a braided metal shield around the wire insulation to intercept any radio interference and carry it to ground (Fig. 4-1). This shield is encased in a tough plastic outer insulator to protect it from abrasion. The conductor of the Slick ignition leads, instead of using stranded wire, uses a continuous spiral of wire and impregnates it with a silicone rubber insulation.

Ignition leads may be of either the 7-mm or 5-mm size, with the smaller being the more common size, by far, for current production harnesses. Some of the older engines use a separate shielding, complete with all of the elbows and nuts. The nuts may be loosened and unshielded wire pulled into the harness for lead replacement.

It is imperative when installing an ignition lead that there be no strain placed on the wire where it enters the terminal end of the spark plug. To prevent any strain, some manufacturers make elbows with several different angles, specifically 70°, 90°, 110° and 135°, while others make their harnesses with only one angle and depend on the flexibility of the wire to prevent any strain. One popular harness uses no elbow but when it is required to make a sharp bend, a bracket is used to hold the lead with a sufficient bend radius that it will not be damaged.

Spark plugs have either a 3/4-20 or 5/8-24 terminal end, and magneto harnesses are made to fit both types of spark plugs (Fig. 4-2).

Figure 4-1.
Typical ignition lead.

Some of the older harnesses used a phenolic or ceramic tube with a coil spring at its end for the terminal connection in the spark plug; these are called cigarettes. The insulation was cut from the stranded conductor far enough back for the wires to stick through the small hole in the end of the terminal, and the ends of the wires are fanned out to provide a good electrical contact and prevent the cigarette from slipping off of the wire. In some instances, a small aircraft nail or pin was slipped through the hole in the terminal into the strands of wire to provide a better connection.

The current practice is to use silicone rubber for the terminal connectors and to crimp the terminal to the wire rather than spreading the strands. The springs usually screw over the end of the terminal and may be replaced if they are damaged.

One of the important considerations when terminating a harness cable is to be sure that the shielding is properly secured. Most manufacturers use a tapered inner and outer ferrule with the shielding between them. The two ferrules are pressed or swaged together, and the shielding then becomes an integral part of the lead, electrically grounded at both ends.

Replacement ignition leads have the spark plug end installed by the manufacturer, but the magneto end is assembled by the A & P after the lead is cut to length.

Figure 4-2.
Typical spark plug
lead terminals.

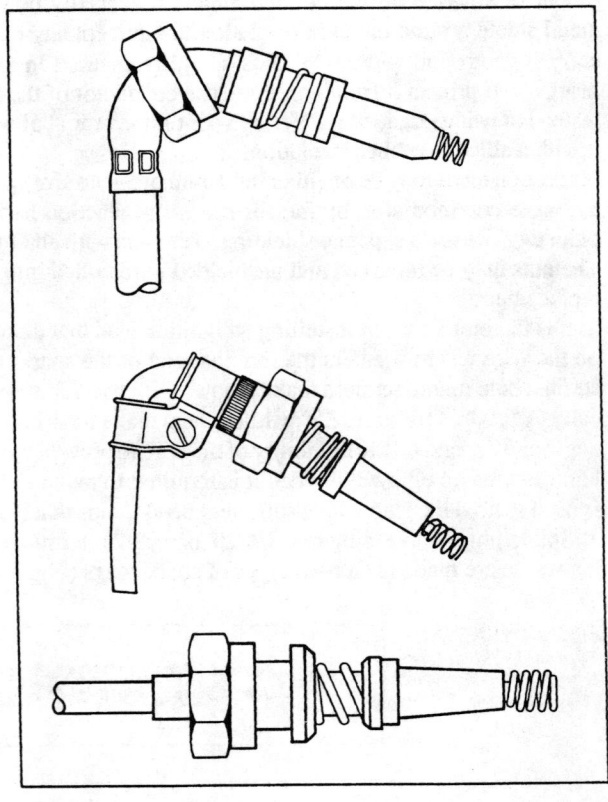

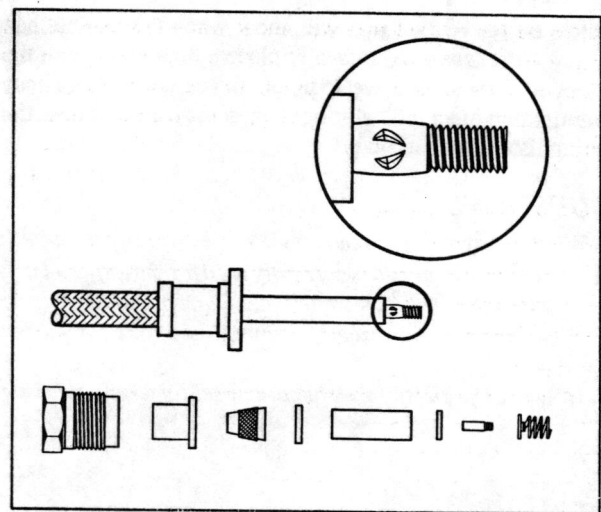

Figure 4-3.
Typical replacement
ignition lead.

If a single spark plug lead becomes damaged, it can be replaced individually without having to replace the entire harness (Fig. 4-3). The various manufacturers have specific instructions which must be followed in detail for terminating an ignition lead, and their service manual must be available—and be adhered to!

B. Testing

Often the ignition problem is blamed on the spark plugs or the magneto, when actually the trouble lies between them. A harness that has a high-resistance leak to ground may bleed off the secondary current before the voltage can rise high enough to fire the spark plug. Often this kind of problem is intermittent because the leakage exists only when the harness is hot. There are a number of harness

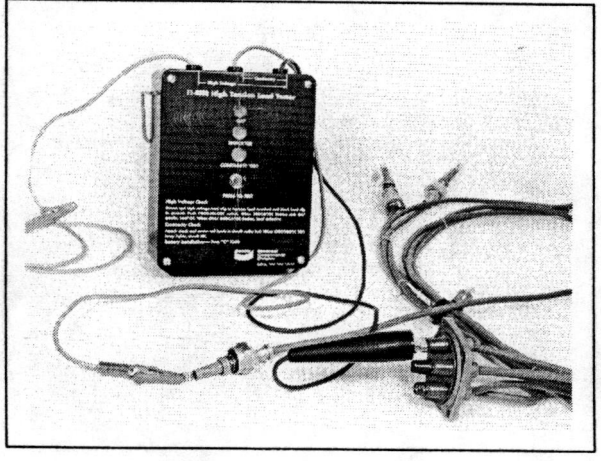

Figure 4-4.
Typical harness
Ignition lead tester.

testers on the market that will show when the harness has excessive leakage (Fig. 4-4). These testers usually place a high voltage on the harness and break it down if there is a weak point. Excessive leakage may be shown by the illumination of an indicator light or, in the case of one of the popular testers, by extinguishing a spark gap.

QUESTIONS:

1. What two types of conductors are used for aircraft ignition leads?
2. What type of material is currently used for the terminal insulator for a shielded spark plug lead?
3. What are the two sizes of terminal end nuts for modern shielded ignition leads?
4. How are the ends of the shielding usually terminated in an ignition harness?

Chapter V
Spark Plugs

A. Spark Plug Nomenclature And Identification

Since the days of the Wright Flyer, spark plugs have been one of the more critical parts of an aircraft engine, yet their seeming simplicity often precludes appreciation of their complexity.

The only function a spark plug has is to provide an insulated electrical terminal in the combustion chamber of a reciprocating engine. To this terminal a high voltage is applied in such a way that a spark will jump to a predetermined ground point and produce enough heat to ignite the fuel-air mixture. This spark plug must resist fouling by any of the contaminants in the combustion chamber. And its electrodes must undergo minimum erosion from either the heat in the cylinder or from the arcing action as the spark jumps between them.

As with all of the components in an airplane, spark plugs have progressed through the process of evolution. The very earliest spark plug was a make-and-break device in the cylinder, producing its spark by mechanically interrupting a flow of current. This was replaced by a spark plug similar to those used in automobiles and motorcycles, in which high voltage from an induction coil punched across an open gap. Then the utility of the airplane was increased by use of two-way radio communications, and shielded ignition systems were developed that contained the radiated energy in a wire braid around the spark plug leads and in a steel shell around the plug itself.

The single ground electrode used by automobiles was improved upon by the use of two, three, or four massive ground electrodes made of a special alloy of nickel. The center electrode evolved from a solid nickel rod to a nickel alloy sheath filled with copper for better heat conduction.

World War II brought out requirements for extended spark plug life, less tendency toward lead fouling with higher lead content fuels, and spark plugs that would resist ice bridging when attempting starts in cold, damp weather. The answer came in the form of fine-wire plugs, having their electrodes made of platinum wire.

1. Size And Shielding

Aircraft spark plugs are made with two sizes of shell threads, 14- and 18-mm; but, with very few exceptions, modern aircraft engines use the 18-mm spark plug.

Although a few of the older engines use unshielded spark plugs, the great majority of engines in service today use shielded spark plugs. Fig. 5-1 shows a short-reach, massive-electrode plug, having low-altitude shielding, or 5/8-24 threads on the terminal end of the barrel. A more watertight seal is provided in the terminal end of a spark plug by recessing the insulator in the spark plug

31

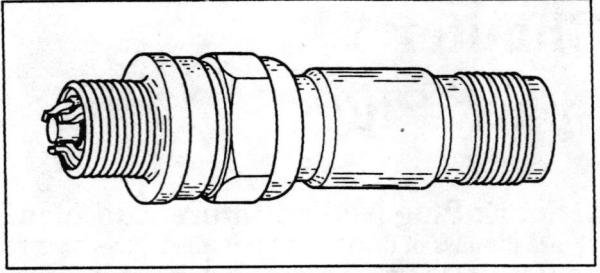

Figure 5-1.
Short-reach,
massive-electrode
shielded spark plug.

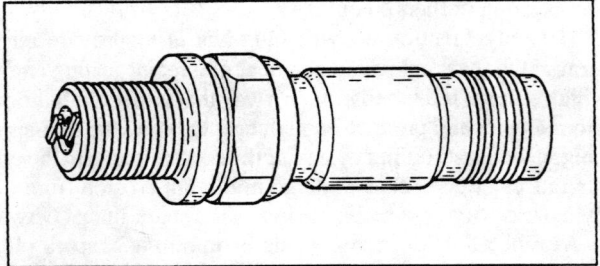

Figure 5-2.
Long-reach, fine-wire
electrode spark plug
with "all weather"
shielding.

shield and using a resilient seal on the harness. This type of spark plug, Fig. 5-2, is called an all-weather or high-altitude spark plug and has a 3/4-20 thread on its terminal end.

2. Electrode

The spark plug in Fig. 5-1 has a three-pronged insert in its firing end, permanently bonded in place, that forms the ground electrodes to which the spark jumps from the center electrode. As the spark jumps it erodes the electrode in much the same way, although on an infinitely smaller scale, as an electric arc welder. The multiple electrodes and large area provide a maximum amount of material, so that the interval between service of these plugs can be extended to the maximum. The center electrode of this type of spark plug as seen in Fig. 5-3 is a nickel sheath, completely filled with copper so there will be a maximum amount of heat transfer from the electrode, and hot spots will not build up in the nickel.

Fine-wire spark plugs, such as seen in Figs. 5-2 and 5-4, have a center electrode and two ground electrodes made of small-cross-section wires of either

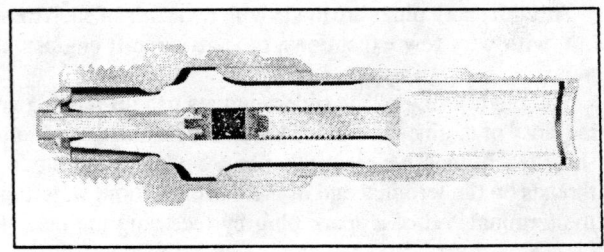

Figure 5-3.
Massive-electrode
spark plug.

Figure 5-4.
Fine-wire electrode.

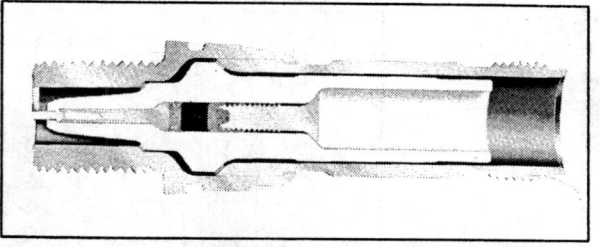

platinum or iridium. The small electrode cross section allows the spark plug to spark at a much lower voltage than the massive electrode, and at the same time the exotic materials of the electrodes resist erosion from both heat and sparking.

3. Reach

The length of the threads on the spark plug shell classifies it according to reach (Fig. 5-5). The shell is threaded for ½ inch on the short-reach spark plug, and for 13/16 inch on the long-reach.

4. Heat Range

The heat range of a spark plug refers to the ability of the insulator and center electrode to conduct heat away from its tip. Fig. 5-6A shows a hot plug, one having a long path for the heat to travel to escape from the tip. Hot spark plugs are used in engines which have a relatively low amount of heat in their combustion chamber—low compression engines. The spark plug in Fig. 5-6B is a cold plug used in a hot-running, high-compression engine. This spark plug has a relatively short path for the flow of heat from the insulator tip to the shell. The spark plug selected for each application must run hot enough to minimize fouling of the insulator tip, while at all times operating at temperatures below that which could cause preignition.

Figure 5-5.
The length of the threads on the firing end of a spark plug determines its reach.

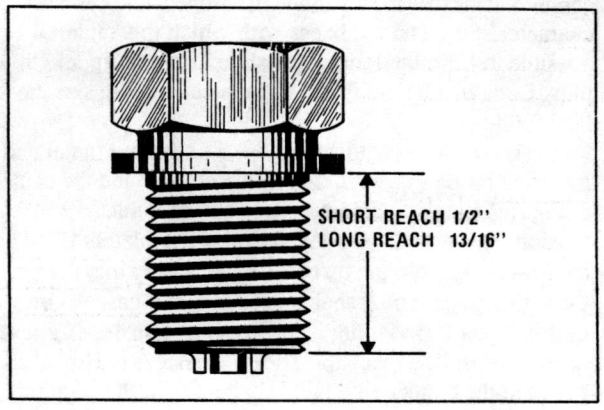

SHORT REACH 1/2"
LONG REACH 13/16"

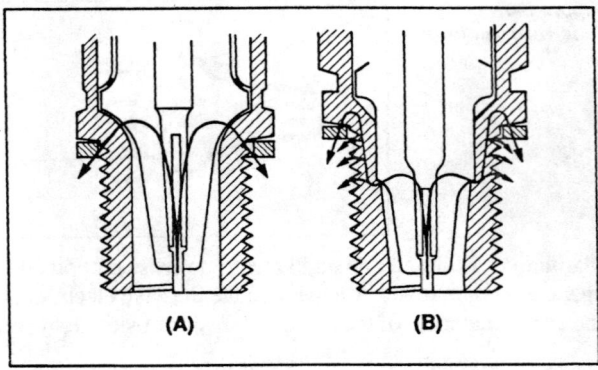

Figure 5-6.
Hot and cold spark plugs.

(A)   (B)

5. Resistors

Shielded ignition, while eliminating the problem of radio interference, causes another problem, that of accelerated electrode erosion. The shield acts as one plate of a capacitor and stores electrical energy. The instant the spark occurs, this capacitor is very rapidly discharged, resulting in high values of current. The discharge makes up what is known as the capacitive component of the spark. And high-amplitude, rapidly changing current is responsible for a large portion of the radiated interference and electrode erosion, but has no effect on engine performance due to its very short duration. A resistor is built into the spark plug to limit the peak current allowed to flow, and thus minimizes electrode erosion.

B. Spark Plug Construction

A special high-strength steel shell and shield are machined with close tolerance to form the holder for the spark plug insulator. After the center insulator is assembled into the shell and the shield, they are bonded together to form an absolutely gas-tight seal. The insulator which holds the contact for the lead wire, the center electrode, and the resistor is made of a special type of ceramic. It is close to a diamond in hardness and has carefully controlled thermal characteristics. The readiness with which this material conducts heat makes it possible to have the longest possible insulator tip length for a given heat range plug. Long insulation tips provide the maximum size cavity and help minimize lead fouling.

The center wire seal is made up of powdered metal and glass or ceramic and forms an intimate bond between the resistor and the center electrode.

Fig. 5-7 shows the center electrode of a massive-electrode spark plug. The erosion-resistant nickel alloy sheath is completely filled with copper to transfer the heat away from the tip of the electrode up into the ceramic insulator and out through the spark plug shell and engine seat gasket. One of the limiting features of this type of spark plug is the low heat transfer efficiency between the center conductor and the insulator. The difference in the linear coefficient of expansion between the copper-filled nickel and that of the ceramic is such that there must

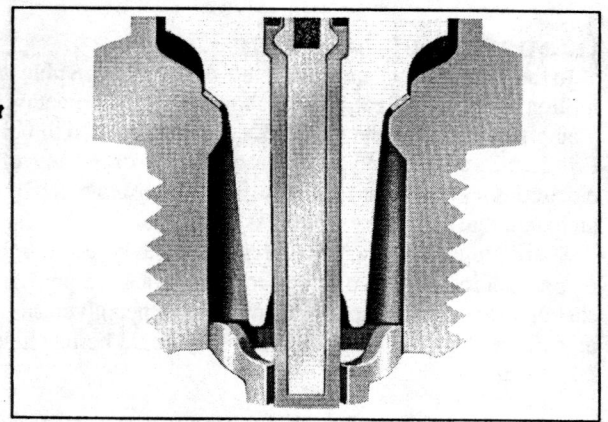

be a slight expansion space between the two. This space forms what may be called a heat dam and prevents a rapid transfer of heat from the electrode into the insulator. This is the main reason for a short insulator tip designed into massive-electrode spark plugs, to keep from causing preignition. The recessed tip configuration used by some manufacturers for their massive electrode spark plugs heats up and cools faster, which allows most of the lead to burn off.

The ground electrode of the massive-electrode spark plug is made of nickel alloy, which is carefully bonded into the end of the shell to assure maximum heat conductivity. The fine-wire plugs have two platinum or iridium wires staked and welded to the rim of the shell. Since the fine wires do not require a mounting ring or any widening of the rim, the cavity in the firing end of the plug is almost completely clear of obstructions (Fig. 5-8). Therefore the gases of the combustion process are free to flow into the cavity and scavenge the combustion by-products before they can condense and form solid deposits.

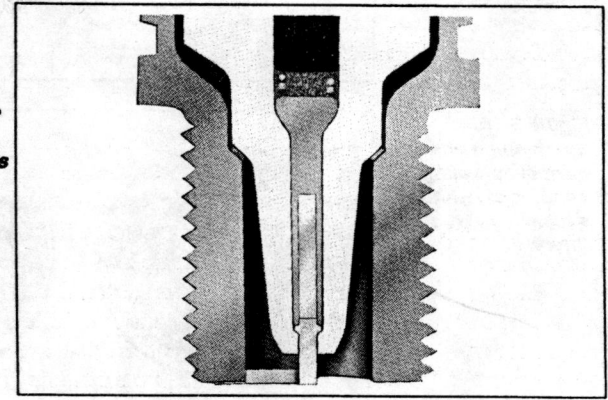

C. Spark Plug Servicing

In the course of one hundred hours of operation, a plug will spark about eight million times and be exposed to about a quart of tetraethyl lead. This is enough arcing to wear away the electrodes and enough lead to build up deposits in the firing end and destroy the heat-conducting capability of the plug. Platinum electrodes used in some fine-wire plugs can be damaged by lead deposits to such an extent that the plug may not be salvageable.

Spark plug servicing is not an operation to be taken lightly, but since proper reconditioning of a set of plugs constitutes a good portion of the time required for an engine inspection, they may be either given less attention than they actually deserve, or replaced when they could be reconditioned at a far lower cost to the owner.

1. Initial Inspection

When spark plugs are removed from an engine for servicing, they should be placed in a rack with numbered holes so you can know from which cylinder and which position they were removed (Fig. 5-9). This will aid you in making an intelligent evaluation of the condition inside the cylinder.

A plug of the correct heat range which has been operating for a short time will have the appearance of the firing ends in Fig. 5-10. The dull brown deposit

Figure 5-9.
When spark plugs are removed from an engine, they should be placed in a rack so they can be identified with the position in the engine from which they were removed.

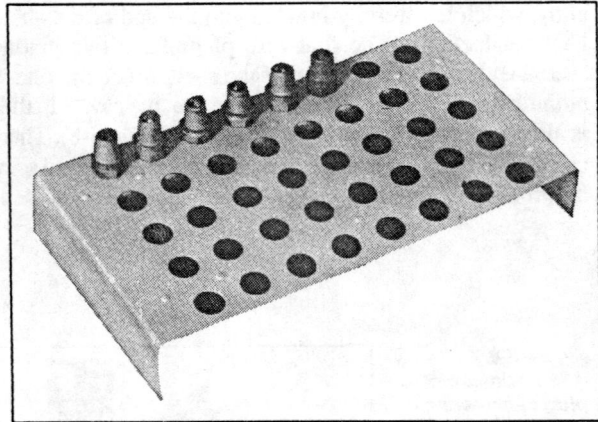

Figure 5-10.
Spark plugs under normal operating conditions, without excessive service intervals.

on the insulator and the lack of buildup in the cavity indicate that the cylinder has been firing properly, and excessive wear has not occurred. The plugs can be cleaned, regapped, tested, and reinstalled for another service period.

Before proceeding any further with spark plug servicing if the initial inspection shows that the plug is worn out, it should be discarded without further expenditure of time. Fine-wire plugs can normally be used until either the ground or center electrode wears away to one-half of its original dimensions. Fig. 5-11 shows a plug worn well beyond its usable life. Massive-electrode plugs can usually be regapped if they have not worn away more than one-half the thickness of the nickel sheath. This sheath is about 0.030 thick and can wear to about 0.015 before the plug will be discarded.

If a few of the spark plugs in an engine appear severely eroded, there is a likelihood that there is a real problem in the cylinders from which these plugs were removed. Spark plugs such as the ones in Fig. 5-12 could indicate an induction leak or other fuel metering malfunction. Check for indications of further damage by using a borescope in the cylinder. Look closely at the piston head and all around the cylinder head. Excessive overheating of a massive-electrode plug can erode all the way through the nickel sheath and allow the copper core to run out.

All aviation gasoline uses tetraethyl lead to improve its anti-detonation characteristics, and ethylene dibromide is added to the fuel to help scavenge the lead oxides before they solidify in the combustion chamber. If the fuel-air mixture has been non-uniform or if the engine or spark plug has been operating

Figure 5-11.
Plugs worn beyond usable life.

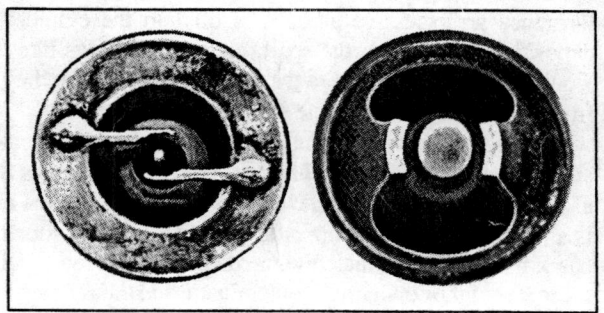

Figure 5-12.
Severe erosion of the electrodes can indicate a condition of over-heating.

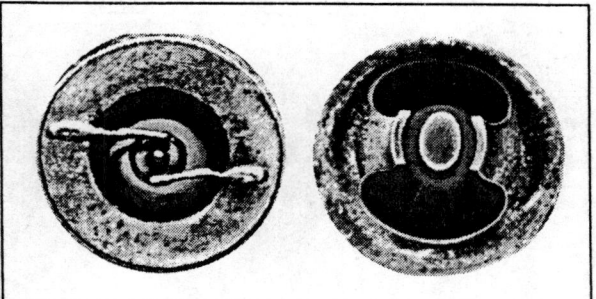

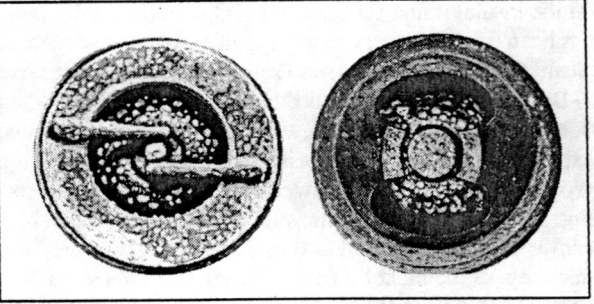

**Figure 5-13.
Lead fouling from
unscavenged
deposits of tetraethyl
lead.**

too cold, lead deposits will form in the firing end cavity of the spark plug and give the appearance of hard, gray, clinker-like deposits (Fig. 5-13).

These must be removed, preferably with a vibrator type cleaner. After a very light abrasive blast, the plug may be regapped, tested, and returned to service.

A black, dry, sooty deposit in the firing end of a spark plug comes from a too rich fuel/air mixture. Too much ground idling with its attendant poor fuel/air mixture distribution or an operation with too rich an idle mixture will cause this, as well as the spark plug having the incorrect heat range for the engine. If the spark plug is correct for the engine, spark plugs like those in Fig. 5-14 may be cleaned, regapped, tested, and reinstalled. These sooty deposits, if dry and not glazed, are usually rather easy to remove with a light abrasive blast.

If the black deposit in the firing end of the spark plugs is oily, there is a very good likelihood that there is damage in the engine. Worn valve guides, broken, worn or misaligned piston rings, or leaking supercharger impeller seals can introduce an excessive amount of oil into the cylinder where it burns and deposits its residue on the spark plug. Spark plugs like the ones in Fig. 5-15 would almost surely require the cylinder from which they came to be removed from the engine to determine the source of the oil.

A hard glaze on the nose insulator of a spark plug indicates that there has been silicon in the combustion chamber. This usually comes from sand and almost always indicates a leak in the induction system after the air filter. Silicon glaze is an insulator at low temperatures but becomes conductive as it gets hot. If a spark plug is contaminated with silicon, it is extremely difficult to remove all the traces of the conductive contaminant, and though the plug checks good while

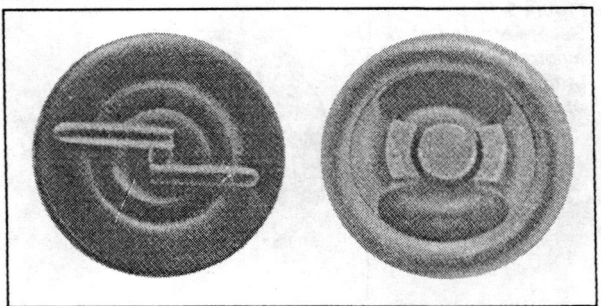

**Figure 5-14.
Dry soot usually
indicates overly rich
operation of the
engine.**

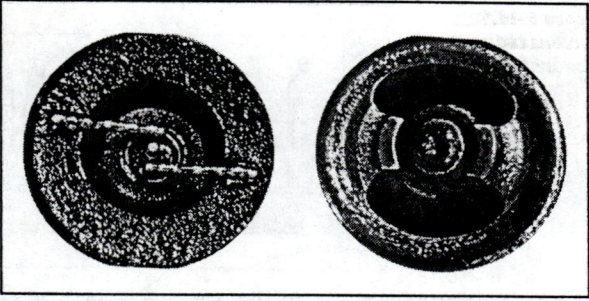

cold, it will foul out when it is hot. It is best to discard any plugs that have a glaze on the nose insulator.

2. Cleaning

It is customary to remove, clean, gap, test, and reinstall spark plugs every one hundred hours, but when this practice is actually analyzed, this may not prove to be the best interval for spark plug service. The two natural attritional forces that affect spark plug life are gap growth and fouling. Both factors are variable, determined by operating conditions, and the life of a modern fine-wire spark plug may be shortened more by unnecessary servicing than by many hours of operation. Plugs should never be cleaned unless they actually need it.

Pilots should be educated in the proper application of throttle for power increase and the proper way to decrease power for descent. Smooth power changes prevent the abrupt temperature changes that are so damaging to spark plug insulators. They should be taught the fine points of the use of carburetor heat which provides better fuel vaporization and helps eliminate lead buildup in the spark plug firing cavity.

To establish a meaningful spark plug service interval, a record should be kept of any ignition incidents, and the condition of the spark plugs should be noted on 100 hour and annual inspections; more often if the engine manufacturer recommends it.

If, when the spark plugs are removed, they have a serviceable gap and no excessive lead deposits, they may be reinstalled for another service interval. When reinstalling the spark plug, be sure that all of the threads are clean. Rotate the spark plugs to the next cylinder in the firing order and change position from top to bottom and bottom to top. Always use new gaskets and the torque called for by the engine manufacturer.

When the buildup of lead deposits in the spark plug indicates the need for servicing, it should be given all of the care a piece of precision equipment deserves.

After the spark plugs have been removed from the engine, degrease them by either soaking them, electrode end down, in a container of Varsol, Stoddard solvent, or unleaded gasoline (Fig. 5-16). Or spray them clean with a suitable solvent, being careful that you don't contaminate the terminal end.

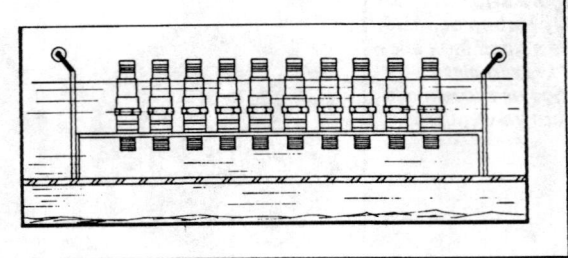

Figure 5-16.
Solvent rack used for loosening deposits.

Other more sophisticated methods are available and can be used in shops equipped with such facilities as vapor degreasers.

Next, remove all of the contaminants from the threads on both ends of the plug by carefully brushing them with a bristle brush or fine-wire brush.

Inspect the threads and clean up any that have been damaged, using a fine triangular file. Now you are ready to break out the lead deposits with a vibrator cleaner.

A bench-mounted cleaner is ideal for removing the lead, but a hand-held vibrating tool will work beautifully (Figs. 5-17, 5-18). A hand pick, ground down from a hacksaw blade, can be used to remove the lead; but the time required makes this a costly way of servicing plugs and also can cause insulator and electrode damage.

After all of the lead deposits have been broken out, the firing cavity can be lightly cleaned with an abrasive blast (Fig. 5-19). The importance of the *absolute minimum* abrasive blasting cannot be over-emphasized. Just as little as five seconds direct abrasive blast on the electrodes could wear as much as two or three hundred hours of engine operation. You must rotate the spark plug while it is being cleaned, in the manner directed by the manufacturer of the cleaning machine, to keep electrode wear to a minimum.

Use only glass beads or aluminum oxide as an abrasive, never silica sand. The abrasive should be fresh and clean; replace it after cleaning about 75 to 100 spark plugs. A water trap should be installed in the air line just before the cleaner. If the cleaner is too full of abrasive it will not do an adequate job of cleaning

Figure 5-17.
Hand-held vibrator cleaning tool.

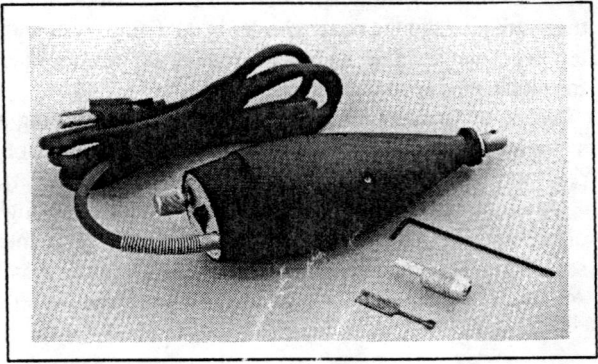

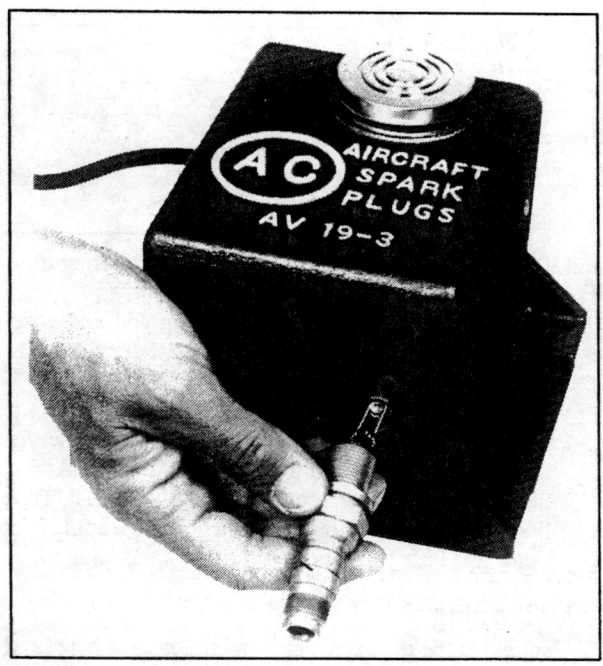

*Figure 5-18.
Bench-mounted
vibrator cleaner,
efficient for cleaning
large quantities of
spark plugs.*

*Figure 5-19.
Cleaning residue is
removed by very
light abrasive
blasting.*

Figure 5-20.
Never clean the
terminal well of a
spark plug with
leaded gasoline.

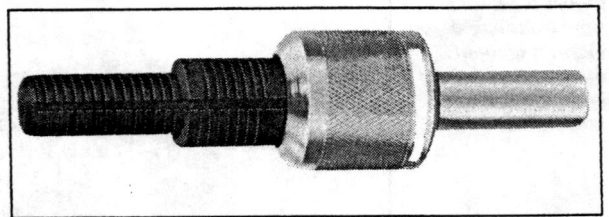

and if the abrasive is worn, it will take an extra long time to remove all of the traces of loosened deposits.

Now is a good time to check the firing end and nose ceramic for general condition. Any unusual appearance is cause for further investigation and possibly a detailed cylinder inspection with a borescope.

Visually check the terminal well for any form of contaminant which could cause the spark plug to misfire. When the plug is serviced, this end must be cleaned with a special cleaning tool or a little Bon Ami on a swab, and a final cleaning with alcohol or acetone (Fig. 5-20). CAUTION: Never use leaded gasoline to clean the terminal end of a plug, as it could become a good electrical conductor at a later time and again cause spark plug misfiring.

3. Gapping

Gapping is the next scheduled operation, and here is where the correct tools *must* be used and the proper procedure followed in detail. Definite guidelines are called out in the spark plug service manuals. If they are followed, prolonged spark plug life can be realized.

Center electrodes in fine-wire and massive-electrode spark plugs are never to be adjusted, moved, or tampered with, as this can cause nose ceramic stress and more serious complications if the spark plug is reinstalled. When adjusting the gap clearance on a massive-electrode spark plug, a special tool should be used to move the ground electrode over to form the correct gap.

During the adjustment of the ground electrodes be sure that the gap is within the tolerance allowed by the engine manufacturer. Move the ground electrode over with the proper tool, being sure there is no feeler wire in the gap area while the gap is being closed. Special caution must be used when making this adjustment, as it is not recommended to open the gap if it has been closed too

Figure 5-21.
Gapping tool for
massive-electrode
spark plug.

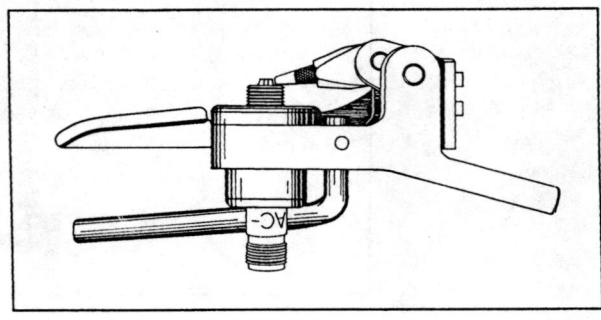

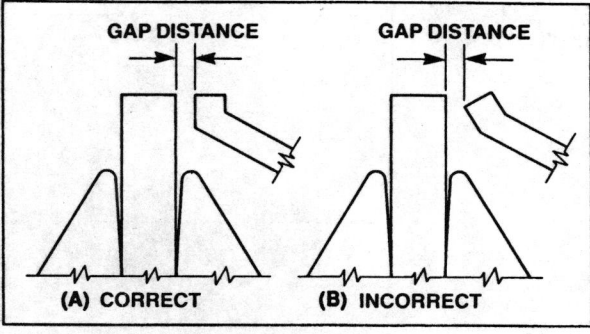

Figure 5-22.
The ground electrode must be moved so that its edge is parallel to the center electrode.

GAP DISTANCE   GAP DISTANCE

(A) CORRECT   **(B) INCORRECT**

Figure 5-23.
Wire gauge used to measure gap distance.

GAP GAGE

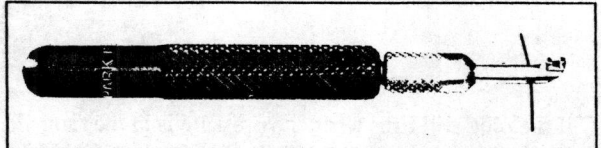

Figure 5-24.
Fine-wire spark plug gapping tool.

tight; nose ceramic or electrode damage could result. The ground electrode must be moved so that its edge is *parallel* to the center electrode. If it is not, the gap will be formed at the electrode edge which will soon erode away and widen the gap (Fig. 5-22). The proper gap distance is measured with a wire-type gauge, never a flat feeler gauge (Fig. 5-23).

Fine-wire spark plugs lend themselves to easy gap adjustment with the use of an inexpensive tool (Fig. 5-24). Caution must be exercised when closing the gap on these spark plugs, as the ground electrodes, especially the iridium electrodes, are quite brittle and can fracture if improperly handled.

4. Testing

Electrical testing consists of assuring yourself that the spark plug will spark in the gap under pressure. The most accurate test available is done with a bomb tester such as the one in Fig. 5-25. Here the plug is placed in the holder, 200 psi of air or nitrogen is introduced into the test chamber and a high voltage applied to the plug terminal. If the spark plug fires consistently, it is considered satisfactory for service. A more commonly used test is performed in the combination cleaner-tester which uses a lower air pressure and, consequently, a lower voltage.

**Figure 5-25.
Bomb tester.**

If the plug still fires when the pressure is in the range identified by "OK" on the gauge, the plug is acceptable for service; but if it blacks out or is intermittent with any lower pressure, the plug should be replaced.

If the spark plugs are not to be used within a day after they have been cleaned, gapped, and tested, they should be stored in a dry cabinet. Any type of closed container, away from contaminants and heated by a 100-watt light bulb, will keep the moisture out sufficiently to prevent rusting. A small amount of non-conductive rust-preventive oil may be applied to the threads of the firing end to protect them and to serve as an anti-seize compound when reinstalling the plug.

Radial engines with odd numbers of cylinders allow every other spark for each cylinder to have opposite polarity. Since the sparking action is similar to an arc welder, this opposite polarity sparking distributes the wear between the center electrode and the ground electrodes. Horizontally-opposed engines have even numbers of cylinders and either two- or four-lobe cams, so each cylinder sparks with the same polarity every time, and this concentrates the wear. Every time the spark plugs are removed from an engine, they should be rotated from top to bottom, and to the next cylinder in the order of firing (Fig. 5-26).

5. Installation

Before installing the spark plug in the cylinder, be sure the threads are perfectly clean. Bronze and steel bushings may be cleaned out by running a special thread-cleaning tool through them.

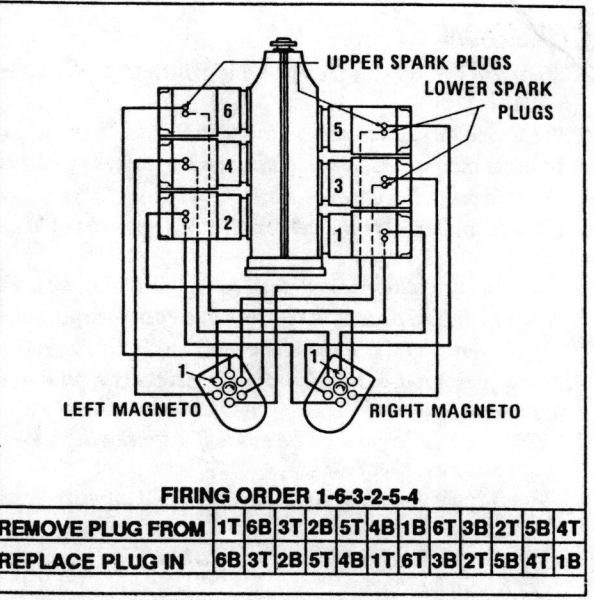

Figure 5-26.
Spark plug
replacement rotation.

FIRING ORDER 1-6-3-2-5-4

REMOVE PLUG FROM	1T	6B	3T	2B	5T	4B	1B	6T	3B	2T	5B	4T
REPLACE PLUG IN	6B	3T	2B	5T	4B	1T	6T	3B	2T	5B	4T	1B

If the grooves in the tool are filled with grease, the powdery residue from the threads will stick in it rather than fall into the cylinder. A thread cleaner should not be used in a heli-coil insert, as there is a danger of the tang of the heli-coil catching in the slots of the tool. But a bottle brush can be used to remove the contaminants from these threads.

Always slip a new gasket over the threads of the spark plug and then turn it into the plug hole. There is no need to use any antiseize compound other than the rust-preventive oil that is already on a new spark plug, or applied during spark plug servicing. The spark plug must screw all the way down to where the gasket makes contact while using only your fingers, in order to get the proper seat with the specified torque. After the plug is all the way down, use a spark plug socket and torque wrench to get the torque recommended by the engine manufacturer.

The required torque value assumes clean threads and the use of a lubricant or anti-rust oil recommended by the manufacturer. This will produce about 0.002-inch compression of the gasket and a good spark plug seal. If too much torque is applied, there is a possibility that the threads may be overstressed, the shell stretched, and the plug damaged. The main danger, however, is not too much but too little torque. With too little torque, hot gases can leak out past the threads and damage both the spark plug and the threads in the cylinder.

Be sure the lead terminal is absolutely clean and insert it straight in to prevent damage; then tighten the lead nut down snugly.

QUESTIONS:

1. What are two advantages of a fine-wire spark plug over a massive-electrode plug?
2. What size thread is used on the firing end of most aircraft spark plugs?
3. What is the purpose of shielding on a spark plug and ignition lead?
4. What is an "all weather" spark plug?
5. Of what material is the center electrode constructed on a massive-electrode spark plug?
6. On what type engine would a "hot" spark plug be used?
7. What is the purpose of a resistor in a resistor spark plug?
8. Of what material are the electrodes made for fine-wire spark plugs?
9. What is a good indication of excessive wear on a massive-electrode spark plug?
10. What inspection should be made if a spark plug shows signs of having been severely overheated?
11. What is the best way to remove the lead deposits from the firing end cavity of a spark plug?
12. What could likely be indicated by a black, oily deposit in the firing end cavity of a spark plug?
13. What are the two natural attritional forces which determine the operational life of a spark plug?
14. Why is it recommended that spark plugs be rotated from top to bottom, and to the next cylinder in firing order when they are replaced after being serviced?
15. How much abrasive blasting should a spark plug be given after all of the lead deposits have been loosened?
16. What can properly be used to clean the terminal end of a shielded spark plug?
17. What should be done with a massive-electrode spark plug that has had the electrode gap closed too much?
18. What special precautions should be used when gapping a fine-wire spark plug?
19. What would be indicated by a spark plug that occasionally "blacks out" when it is being tested in a combination cleaner-tester?
20. What is the danger of installing a spark plug with too much torque?
21. What is the danger of installing a spark plug with too little torque?

SECTION 2:
ELECTRICAL POWER SYSTEMS

Chapter VI
DC Generators

A. Electromagnetic Generation

1. Left-hand Rule

As we reviewed at the beginning of this book, any time a conductor is moved in a magnetic field, it cuts across the lines of flux and a voltage is generated in the conductor; this voltage causes a current to flow. In Fig. 6-1A, when the conductor is moved out from the magnet, the lines of flux will encircle the conductor as shown and, according to the left-hand rule for generators, *if the fingers of the left hand encircle a conductor in the direction of the lines of magnetic flux, the electron flow will be in the direction the thumb is pointing.*

2. Simple AC Generator

A single conductor in a magnetic field is not an efficient producer of electricity, so to improve the output, let's consider the conductor to be formed into a loop with ends attached to slip rings (Fig. 6-1) with brushes connecting the rings to the load. Now if this loop is rotated in a clockwise direction, the side *B* of the loop in Fig. 6-1A will move downward as it passes in front of the north pole of the magnet. And, according to the left-hand rule, the lines of flux will encircle the conductor in such a direction as to cause the electrons to flow toward the slip rings; the load current will be in the direction shown by the arrows.

Side *A* of the coil is, in the meantime, moving upward, across the south pole of the magnet, and the direction of the lines of flux causes the electrons to flow away from the slip rings. This flow gradually decreases until, in the position shown in Fig. 6-1B, both conductors are traveling parallel to the lines of flux and no lines are being cut; thus there will be no voltage generated, and no current will flow.

Now, in Fig. 6-1C, the conductors are again moving across the face of the magnet, and the maximum number of lines is being cut. A voltage is again generated in the wire, but this time the polarity is opposite to that in Fig. 6-1A, and the electrons flow in the opposite direction. The output of this generator is

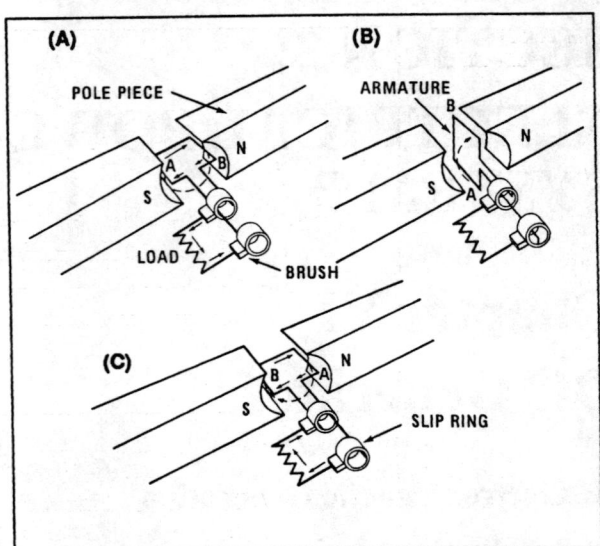

Figure 6-1.
Principles of an AC generator.

a sine wave; that is, one whose amplitude varies as the sine of the angle through which the coil has rotated.

The amount of current produced in generators such as this is dependent on three things: the number of lines of magnetic flux, the number of conductors cutting the flux, and the speed at which the conductors move through the flux. In essence, the amount of current depends on the rate at which the lines of flux are cut.

3. Simple DC Generator

Rather than attaching the ends of the moving coil to slip rings, we can produce direct current by attaching the ends to a split ring. This must be done in such a

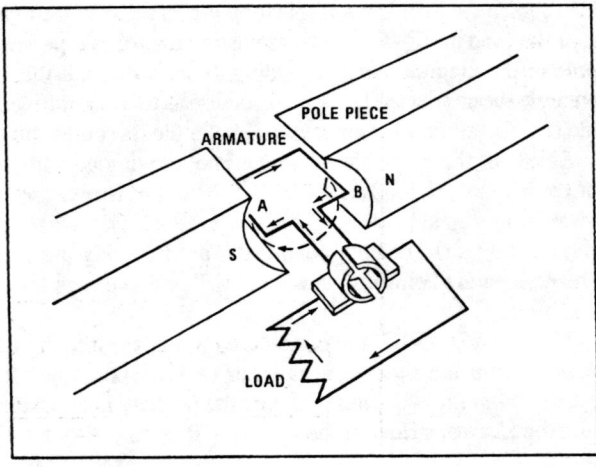

Figure 6-2.
Principles of a DC generator.

48

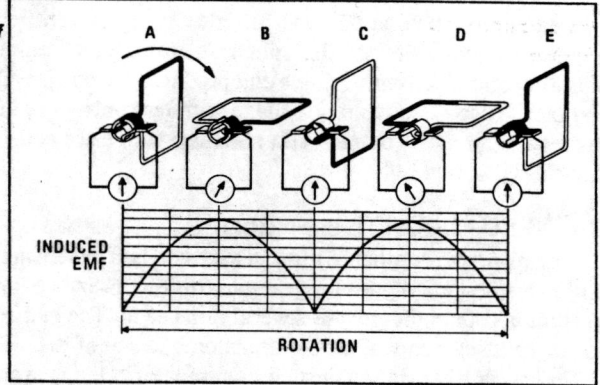

Figure 6-3.
The output voltage of a DC generator is pulsating direct current.

way that the side of the coil passing in front of the north pole, Fig. 6-2, will always be connected to the negative brush, and the side moving across the south pole will feed the positive brush. AC is generated in the armature coil, but the split-ring commutator causes electron flow through the load to be in the same direction at all times.

Fig. 6-3 shows the output of a single-coil DC generator. In position A, the conductors are moving parallel to the magnetic field, and no lines are being cut, so no voltage is being generated. In position B, the black side of the coil is moving downward across one of the poles, and the white side moves upward across the opposite pole. Maximum voltage is generated and maximum current flows through the load. In position C, the conductors are again moving parallel to the field and no current is generated. In D, the white side of the coil is moving downward and the black side upward. The flow in the conductor is opposite to that in position B, but since the white side is now feeding the black brush, the flow in the load circuit is the same as it was in B.

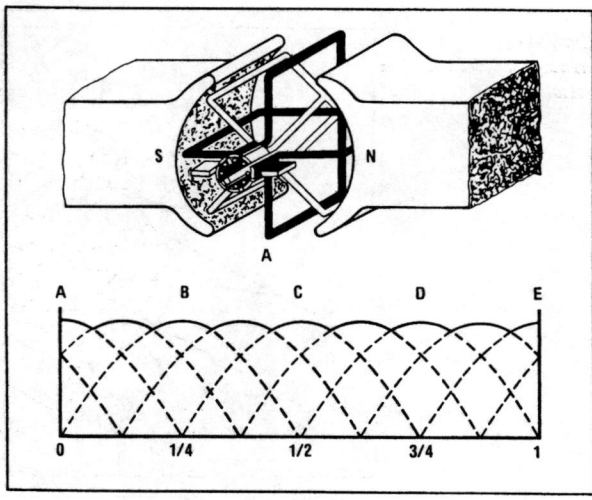

Figure 6-4.
The use of more than one armature coil produces a smoother DC output.

49

A generator such as this one has alternating current in the conductor, but because of the action of the split ring, called a commutator, the output is pulsating direct current. A single coil produces the pulsations shown in Fig. 6-3, and in order to get an output voltage with less pulsation, more coils are added to the armature. The output of an armature with three coils would be similar to that shown in Fig. 6-4.

4. Multi-coil DC Generator

Increasing the number of turns in each coil and increasing the number of coils will increase the output of a generator. In Fig. 6-5A we have a schematic of a generator with eight coils of several turns each. The ends of these coils attach to adjacent segments of the commutator, one end of two coils to each segment.

Carbon brushes ride against the commutator, taking the current from the coils. In Fig. 6-5A, we see that the brushes short-circuit two of the coils during the time they are moving parallel to the magnetic field, and in this position there is

Figure 6-5.
Lap-wound armature coils.

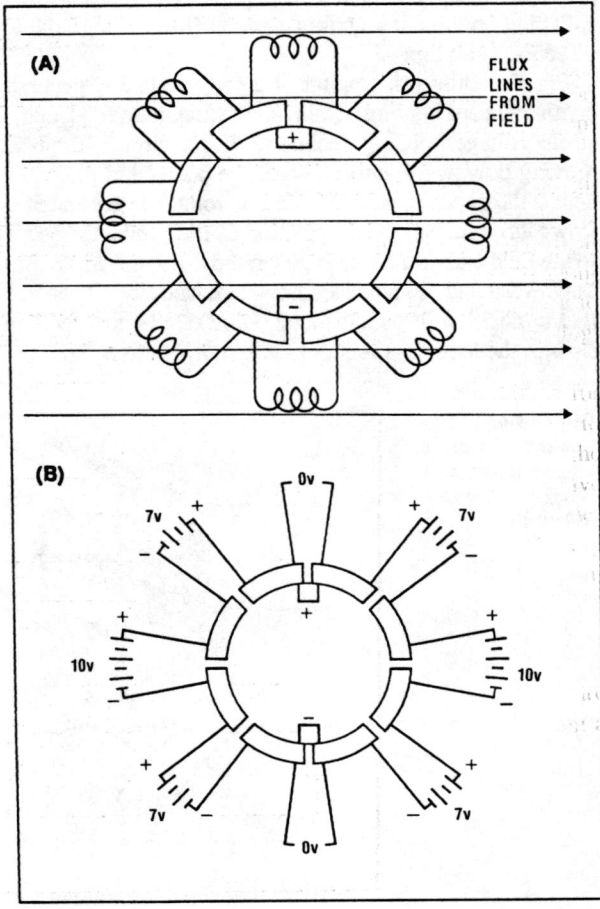

a minimum of voltage difference between the two segments of the commutator, and brush arcing will be minimum.

As the armature rotates, the coils progressively cut across more lines of flux from the field, and the coils that cut the most flux have the greatest voltage in them. Fig. 6-5B is an analogy of the voltage in the coils. Across the segments where the brushes ride there is zero voltage, or at least a minimum of voltage. When the coil has rotated 45 degrees, about 70 percent of its voltage will be produced, and at 90 degrees the voltage will have risen to peak which, in the example, we are calling 10 volts. As the armature continues to rotate, the voltage in that coil drops until it is under the opposite brush, at which time there is zero voltage across it.

5. Armature Reaction And Its Control

In Fig. 6-5, we saw that a brush contacting the coils traveling parallel to the field flux will find a minimum voltage difference between the adjacent commutator segments. This condition could exist only if there was no current flowing in the coils of the armature; but, as we know, when armature current does flow, magnetic flux will surround the coils and deflect the lines of the field flux.

Figure 6-6A: With no current flowing in the armature coils, the neutral plane (a line perpendicular to the lines of flux) lies directly over the brushes.

Figure 6-6B: Current flowing in the armature coils produces a magnetic field around each winding of the coil.

Figure 6-6C: The reaction of the two fields shifts the neutral plane away from the brushes. There will be a voltage between the adjacent commutator segments, and arcing will occur at the brushes.

To summarize Figure 6-6, the neutral plane perpendicular to the flux lines will be shifted from alignment with the brushes, so that there is now a potential

Figure 6-6.
Armature reaction.

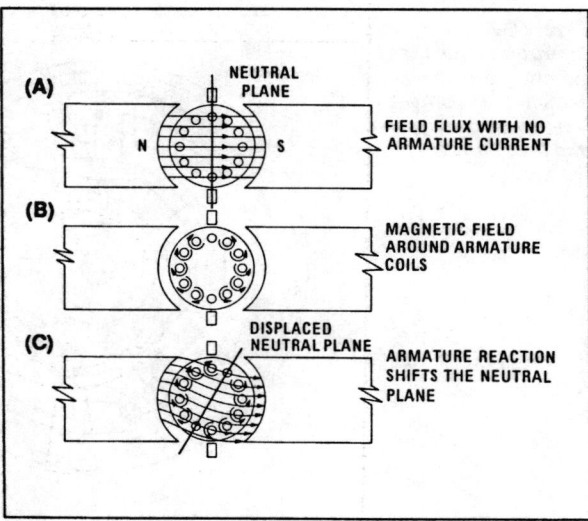

51

difference, or a voltage, between the adjacent segments; and when the brush shorts them out, arcing will occur.

This displacement of the neutral plane is called armature reaction, and it varies with the amount of current flowing in the armature, which is, of course, the load current.

One of the simpler provisions for minimizing the effect of armature reaction is to position the brushes in such a way that they will be in the neutral plane when the rated current of the generator is flowing. Some generators have a movable brush rigging so the brushes can be moved, while the generator is on the test stand, to a position which produces minimum arcing when the specified current is flowing.

Shifting the brushes to minimize the effect of armature reaction is not efficient, since the neutral plane shifts as the load current changes; so in order for the neutral plane to remain aligned with the brushes regardless of the load current, interpoles such as those seen in Fig. 6-7 may be used.

An interpole is placed between each of the main pole pieces and wound so that its polarity is the same as that of the next main pole in the direction of rotation. As the field of the armature winding tends to shift the neutral plane, the field of the interpole, whose winding is in series with the armature and whose strength also varies with the load, counteracts the armature reaction and holds the neutral plane aligned with the brushes.

Generators which produce outputs of several hundred amperes require more compensation than can be provided by interpoles alone, and compensating windings are used on these to minimize brush arcing.

Fig. 6-8 illustrates the way arcless commutation is achieved on a 200-amp, eight-pole DC generator. The high current from the negative brushes flows through the windings of the interpoles and also through coils imbedded in the faces of the main poles. This winding increases the effectiveness of the inter-

Figure 6-7.
Interpoles cause the neutral plane to remain in alignment with the brushes as the load current varies.

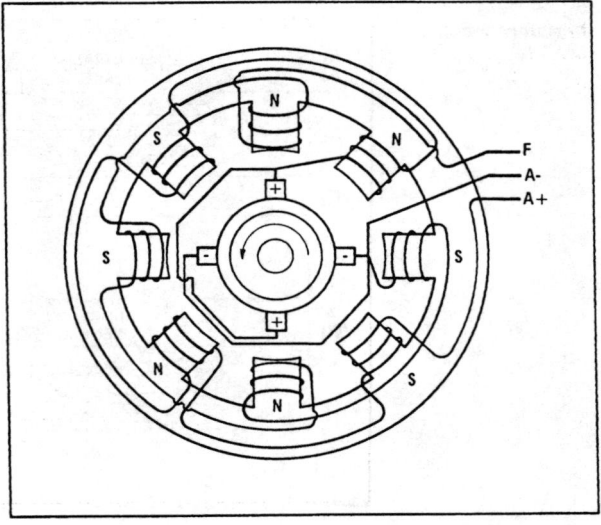

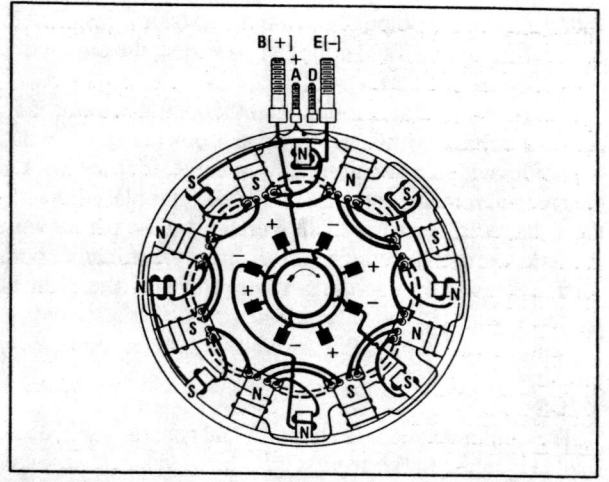

Figure 6-8. High-output generators use compensating windings in series and provide arcless commutation.

poles and holds the neutral plane in its proper position, even with wide variations in output current.

6. Generator Field Windings

a. Shunt

The voltage generated in an electromagnetic generator is determined by the *rate* at which the lines of flux are cut. This rate is a function of the number of conductors, which is fixed by the construction of the generator, by the speed or rotation—the RPM—of the engine, or by the strength of the magnetic field. Practical voltage control is accomplished by varying the strength of the magnetic field. Fig. 6-9A is a schematic of a simple shunt-wound generator, in which the field flux is provided by electromagnets whose coils are in parallel, or shunt,

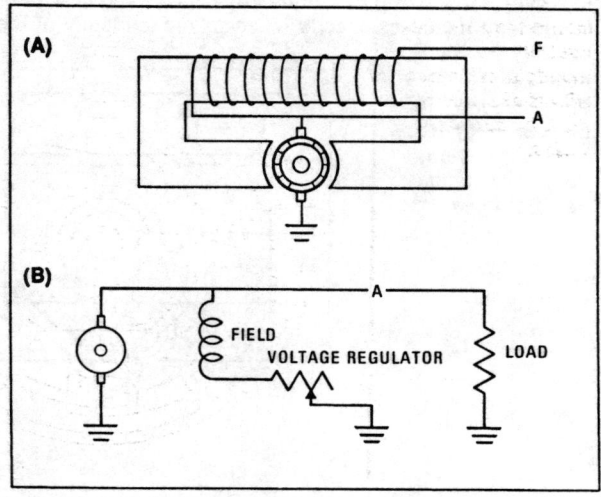

Figure 6-9. Shunt-wound generator.

with the armature output. If a variable resistor is installed in such a way that the current through the field coils can be varied, the output of the generator can be controlled (Fig. 6-9B).

One of the limitations of a shunt-wound generator is its inability to hold a constant voltage with a varying load. Look at Fig. 6-9B and see that the field is in parallel with the armature. Let's assume, for a moment, that the resistance of the rheostat remains constant. When a load is placed in series with the armature, the voltage drop across it will increase and its output voltage will drop. Since the field excitation voltage is taken from the armature output, it will also drop. This voltage drop decreases the strength of the field and causes a further decrease in the output voltage. This voltage variation with load can be prevented by using an automatically variable resistor, a voltage regulator, in the field circuit.

b. Series

The output voltage of a shunt-wound generator will decrease when the output load is increased if no regulation is provided. A series-wound generator, on the other hand, will act in exactly the opposite way. As the load current increases, the strength of the field and the output voltage will both increase (Fig. 6-10).

The difficulty in controlling the output of series-wound generators makes them impractical for use in aircraft power systems. But since their output characteristics are opposite those of shunt-wound generators, both types of windings may be installed in one generator. This produces a compound generator, one whose output voltage is less affected by load variations than that of either of the more simple generators.

c. Compound

High-output generators may have two windings around their poles, one in series with the armature and the load, and one in parallel. Fig. 6-11 shows the way the fields of a compound generator are connected.

The relative strengths of the two magnetic fields determine the characteristics of a compound generator. If the effect of the series field to increase the voltage as the load increases exactly balances the tendency of the shunt field to allow

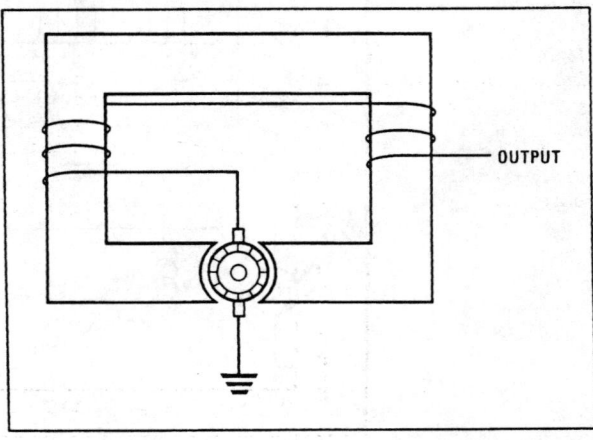

Figure 6-10. Series-wound generator: All of the armature current flows through the field windings.

OUTPUT

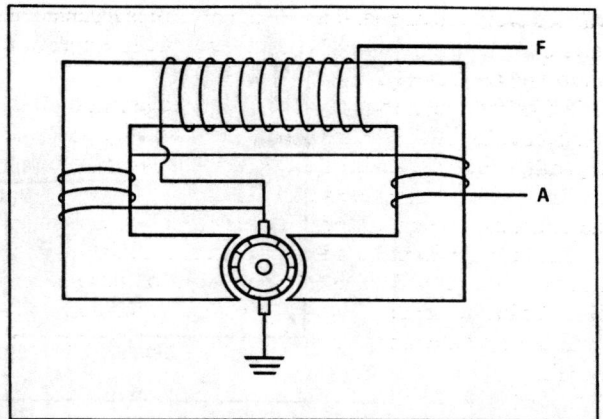

Figure 6-11.
Compound-wound
generator.

the voltage to decrease with the load, the generator is said to be flat-compounded (Fig. 6-12). If the effect of the series field is greater than that of the shunt field, the generator is over-compounded, and if the shunt field allows the voltage to drop at rated current, the generator is under-compounded.

If the generator is over-compounded, the series field is predominant, and the voltage rises with the load. If it is under-compounded, the shunt field has the most effect, and the voltage drops with an increase in load current.

B. Control Of Aircraft DC Generators

1. Controls Required
a. Voltage Control

An aircraft DC generator, as we have just seen, is self-excited; that means that the current used to produce electromagnetism in the field comes from the armature (Fig. 6-6B). If there were no control in the field circuit, the generator would run away with itself and burn out its windings. As the generator starts to

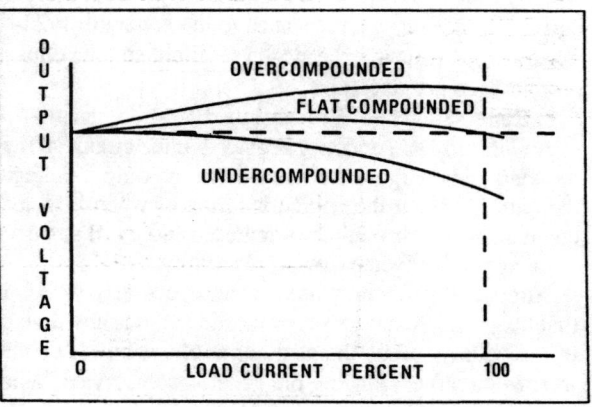

Figure 6-12.
In a flat compounded
generator, the
voltage remains
relatively constant as
the load current
changes.

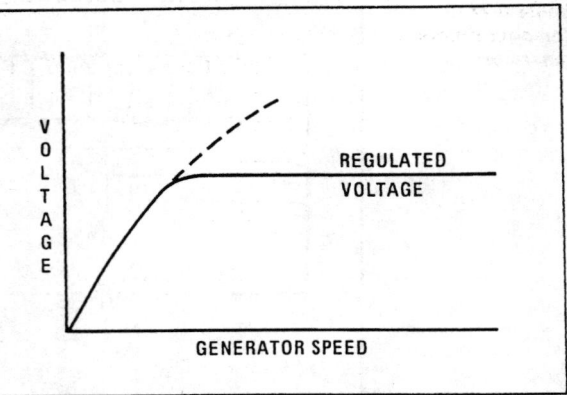

Figure 6-13.
The voltage regulator allows voltage to rise to its regulated value, then holds the output voltage constant.

turn, a voltage begins to build up because of residual magnetism in the field frame. This voltage causes current to flow in the field coils and increases the voltage. As voltage rises, field current increases and, by a bootstrapping action, the voltage will continue to rise as indicated by the dashed curve in Fig. 6-13.

When the voltage required by the system is reached, the voltage regulator, which is actually a variable resistor in the field circuit, will automatically decrease the field current and prevent the voltage rising above the regulated level. The voltage is held relatively constant by controlling the amount of field current allowed to flow.

b. Current Limitation

All aircraft generators are rated with regard to the amount of current they are allowed to produce. Large, high-output generators are prevented from exceeding their rated current output by a special heavy-duty fuse called a current limiter between the generator output and the main bus.

Generator systems having lower outputs, such as are found on many smaller general aviation aircraft, have an automatic current limiter which places a variable resistance in the field circuit any time the output current exceeds the generator rating. This resistor decreases field current and lowers output voltage, which in turn decreases the output current. Both the voltage regulator and the current limiter place a resistance in the generator field—the voltage regulator, when output voltage exceeds the set limit, and the current limiter, when output current becomes excessive.

c. Reverse-current Prevention

An aircraft DC generator is placed in the circuit so it will provide current for the load and charge the battery when the output voltage is higher than that of the battery. When the engine is idling or when it is not operating, the battery could discharge through the armature windings if some provision were not made to prevent this reverse flow of electricity.

Almost all generators use a voltage-operated switch in the generator output that keeps the generator off of the aircraft bus any time its voltage is below that of the battery. When the generator voltage builds up above that of the battery, the switch closes, placing the generator in service. When the generator output

56

drops *below* that of the battery, current flows from the battery into the generator, and this reverse flow neutralizes the magnetic effect which had closed the switch. The contacts open and take the generator off the line.

Alternators are attaining a great deal of popularity as a source of direct current and, as we will see, instead of rectifying their AC output with a commutator and brush arrangement as the generator does, solid-state diodes do the rectifying. These diodes not only change the generated AC into DC, but they prevent any flow of direct current from the battery entering the alternator stator windings.

d. Paralleling Requirements

When two generators feed the electrical system of an airplane, provision must be made to keep the voltage output of the two as near the same as possible, to prevent one generator carrying all of the load while the other carries none.

Carbon-pile voltage regulators, as we will see, have provisions for a load imbalance to automatically increase the voltage of the low generator and at the same time decrease the voltage of the one carrying the greater load. Vibrator and transistor regulators normally depend on their ability to keep the voltages constant once they have been adjusted. The generators and regulators are brought up to operating speed and temperature, with their output voltages adjusted, using a highly accurate voltmeter. Then, with both generators on the line and an appreciable load applied, the regulators are adjusted until the generators produce the same current. This procedure, as with any adjustment, must be done in careful accordance with the manufacturer's recommendations.

2. Vibrator-type Generator Control System

The majority of low-output aircraft generators are regulated with vibrator-type controls. The most familiar regulator has three units in one waterproof, metal-covered housing.

a. Voltage Regulator

The output from the generator armature, entering the regulator unit through terminal *G* (Fig. 6-14) passes through the heavy winding *A* of the current limiter and the current winding *B* of the cut-out relay into its armature. No current can flow to the battery or the load until the points of the reverse-current cut-out relay close. As the generator output voltage rises, the magnetic strength of the voltage coil *E* increases enough to close the contacts and put the generator on the line. Load current flowing through the current coil *B* now aids the voltage coil in holding the contacts tightly closed.

The field coils of the generator are excited from the output of the armature, and the ground circuit of the field is completed through the points of both the voltage regulator and the current limiter. When the voltage rises to its regulated value, the magnetic pull produced by the shunt winding, *C*, and the series, or accelerator, winding, *D*, opens the voltage regulator points. With the points open, the ground circuit of the field now includes the resistor, *R*, and the field current drops, as does the output voltage. You will notice that there are two coils producing the magnetic pull to open the contacts. When the points open, the accelerator winding circuit is opened and its field collapses completely, rapidly

57

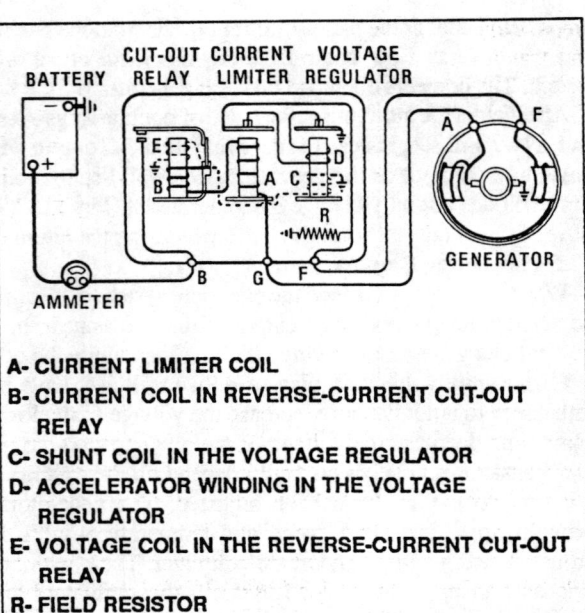

Figure 6-14.
Three-unit vibrator-type generator control, with an accelerator winding on the voltage regulator.

A- CURRENT LIMITER COIL

B- CURRENT COIL IN REVERSE-CURRENT CUT-OUT RELAY

C- SHUNT COIL IN THE VOLTAGE REGULATOR

D- ACCELERATOR WINDING IN THE VOLTAGE REGULATOR

E- VOLTAGE COIL IN THE REVERSE-CURRENT CUT-OUT RELAY

R- FIELD RESISTOR

decreasing the flux, so the spring can close the points more rapidly than if this winding were not used.

Some regulators employ a second resistor in the regulator housing, this one in parallel with the generator field. When the field circuit opens and the field current begins to collapse, the inductance of the coil produces a voltage surge which tends to cause arcing at the points; this resistor suppresses the arcing.

In normal operation, the points in a vibrator-type voltage regulator open and close between 50 and 200 times per second to hold the voltage at a constant value.

In some generators one end of the field is attached to the grounded brush, and a positive voltage is placed on the end of the field in the regulator to control the voltage (Fig. 6-15). This type of system is called a "B" system by the manufacturer, Delco-Remy, to distinguish it from that in which the field is grounded in the regulator, the "A" system.

Before the voltage rises to the regulated amount, current flows through the field from the armature of the cut-off relay through both the voltage regulator and current limiter points and through the field to ground in the generator. When the voltage rises to the value at which the regulator is set, the voltage regulator points open, and resistor R_1 is inserted in the field circuit, decreasing the field current and dropping the output voltage. When the field circuit is opened and the current drops, the induced voltage surge would tend to cause arcing at the points; but since resistor R_2 is in parallel with the field coils, it will shunt off some of the current and minimize the arcing.

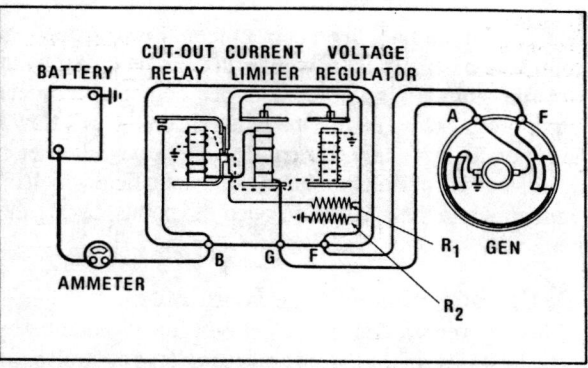

Figure 6-15.
Three-unit
vibrator-type
generator control,
with voltage
regulator between
the positive side of
the armature and the
field.

Some models of vibrator-type voltage regulators have two sets of movable points and one fixed point (Fig. 6-16). When the engine is turning at a relatively slow speed and the field current demands are high, the regulator vibrates between the center and lower contact. When the voltage rises to the regulated amount, the points open and a resistor is inserted into the field circuit. When engine speed is high and field current demands are low, the magnetic pull of the voltage regulator is strong enough to cause the points to vibrate between the center and top contacts. When the contacts are open, the resistor is in the field circuit; but when the voltage is high enough to close the top contacts, the field winding is shorted out and no field current flows.

b. Current Regulation

Any time current taken by the load reaches the rated value of the generator, the magnetic field produced by the heavy winding, *A*, of the current limiter will open the limiter points and insert a resistor into the generator field circuit. This resistance in the field will lower the output voltage and decrease the current. When the current drops, the points close, allowing the voltage to again rise. As long as the demands for current exceed the generator's rating, the current-limiter points will vibrate.

c. Reverse-current Cut-out Relay

When generator voltage rises above that of the battery, the magnetic field of the shunt winding in the cut-out relay will close the points and place the

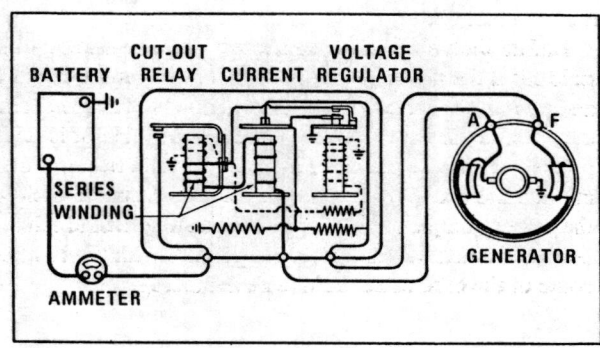

Figure 6-16.
Three-unit
vibrator-type
generator control,
with a double set of
contacts in the
voltage regulator.

generator on the line. Load current then flows through the series winding and produces a magnetic field, which aids that produced by the shunt winding and holds the points tightly closed. When engine speed decreases and generator output drops below that of the battery, current will flow from the battery into the generator armature. Current flowing through the series coil winding of the reverse-current cut-out produces a magnetic field which opposes the field of the shunt coil, and the spring will open the points, taking the generator off of the line.

3. Carbon-pile Voltage Regulator

Voltage is controlled in an electromagnetic generator by controlling the field current, and one of the earlier types of voltage control, one which still finds some application, is the carbon-pile voltage regulator.

Thin discs of pure carbon are stacked inside a ceramic tube (Fig. 6-17) and pressure is applied to the stack by a spring. An electromagnet acts on an armature attached to the spring, so that the total force will be proportional to the output voltage.

Fig. 6-18 is a basic circuit of a carbon-pile voltage regulator used with a compound generator. The electromagnet coil is in parallel with the generator output and in series with both a variable and a fixed resistor. When the output voltage rises, the pull of the electromagnet increases and overcomes some of the spring force, loosening the stack of carbon discs and increasing its resistance. This increased resistance decreases the field current and lowers generator output. As output voltage drops, current through the electromagnet decreases, and so does its pull on the spring which allows it to force the discs tighter together. This decreases the resistance of the pile and increases the generator field current.

Carbon-pile regulators are used with relatively high-output generators and often on multi-engine aircraft. The paralleling system with a carbon-pile regulator maintains a balance between the loads of the generators. This is done by using a paralleling coil on the regulators to increase voltage on the generator

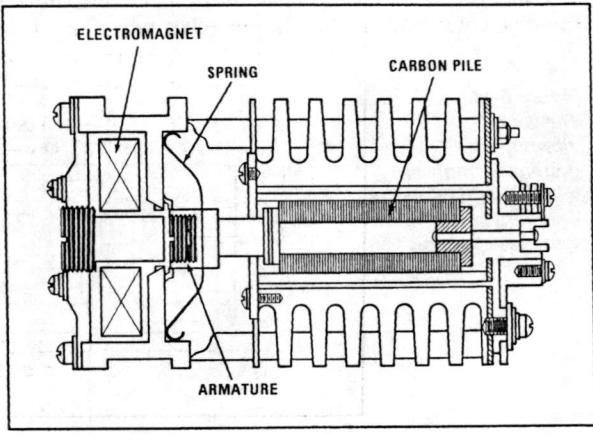

Figure 6-17.
Carbon-pile voltage
regulator.

ELECTROMAGNET

SPRING

CARBON PILE

ARMATURE

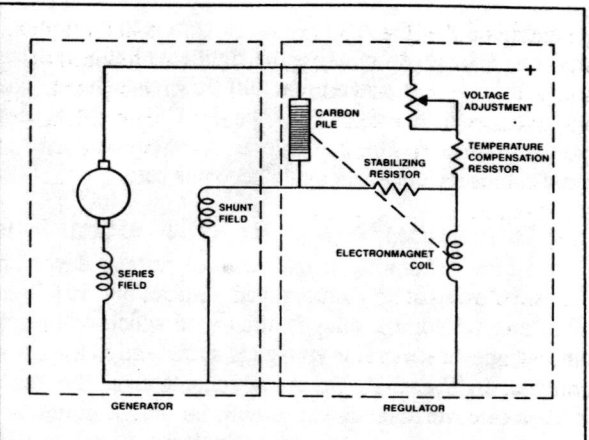

Figure 6-18.
Circuit for generator with a carbon-pile voltage regulator.

carrying less than its share of the load, and to decrease voltage on the one carrying most of the load.

Fig. 6-19 is a schematic of a twin-engine installation using two generators and two carbon-pile regulators. A low-resistance equalizing resistor is installed in the ground lead of each generator. The value of these resistors is such that when the rated generator current flows through them, there will be about a half-volt drop across them.

The voltage developed across the equalizing resistor of the left generator causes current to flow through its equalizing coil, through an equalizing switch, S_E, and to the equalizing coil of the right regulator. From there the current flows to the top of the equalizing resistor on the right generator.

When the load is shared equally between the two generators, there will be no voltage difference across the equalizing circuit, and no current will flow. But

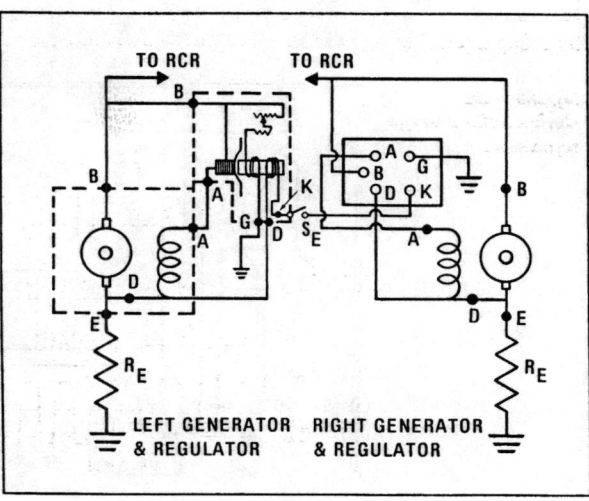

Figure 6-19.
Equalizing circuit for two generators using carbon-pile voltage regulators.

let's assume that the right generator begins to carry more of the load than the left. The voltage drop across the right equalizing resistor will then rise above that across the left, and current will flow through the equalizing coils in such a direction as to decrease the voltage of the right generator and increase the voltage of the left. This continues until the currents are the same, and the voltage drop across the equalizer again becomes zero.

4. Differential Voltage Reverse-current Relay

Most lower-output generator control systems depend on the reverse-current cut-out closing at a predetermined voltage; but with higher-output generators, a differential voltage relay is often used which will place the generator on the line whenever generator voltage is sufficiently high above the *existing* battery voltage. In Fig. 6-20 we see a schematic of the generator connected to a carbon-pile voltage regulator, with its output going to a differential-voltage reverse-current relay, and then to the bus.

The pilot is furnished with a generator switch with which he can disconnect the generator from the line. When this switch is closed and generator voltage rises sufficiently high, the voltage coil closes its contacts, which places the differential voltage coil between the generator output and the main bus. When there is sufficient voltage difference, with the generator being the higher, the differential-voltage coil closes its contacts, allowing current to flow into the main contactor coil, closing it, and placing the generator on the line.

When generator output drops below that of the battery, current will flow from the battery into the generator through the reverse-current coil. This current sets up a magnetic field which causes the differential voltage contacts to open, breaking the circuit to the main contactor coil and taking the generator off of the line.

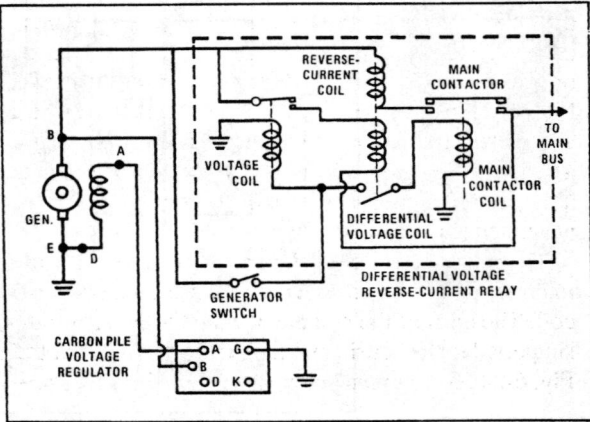

Figure 6-20.
Differential-voltage
reverse-current relay.

62

Figure 6-21.
Typical alternator
used in general
aviation.

C. DC Alternators And Their Control

1. Brush-type Alternators

For many years direct current for use in aircraft was produced in a generator. The armature produced alternating current which was rectified by the brushes and the commutator. The main reason for using this inefficient way of producing electrical energy was the lack of non-mechanical rectifiers of sufficiently small size, which could stand the high temperature and vibration existing in a generator. When semiconductor diodes with high current rating and small size became available, there was a switch to the more efficient alternator.

Now we often see these devices referred to as *AC* generators or *AC* alternators, and they actually do generate AC in their windings; before the AC leaves the housing, however, it is converted into direct current.

The alternator, such as the one shown in Fig. 6-21, has a number of advantages over the generator. In the first place, perhaps one of the most important advantages of the alternator is the fact that load current is generated in the stator or stationary winding and does not have to flow through brushes to the load. Fig. 6-22 illustrates the three-phase stator commonly used in aircraft alternators. The coils in this stator are connected as three windings, joined together to form a Y. There are seven pairs of poles in the rotor and seven coils in each leg of the Y, so, as the rotor turns within the stator, three phases of alternating current are generated (Fig. 6-23).

The rotor of an aircraft alternator is made up of two soft iron end pieces with intermeshing poles pressed onto the shaft, on either side of a drum-type exciter coil. The ends of the coil are attached to two slip rings, and the entire rotor is supported in the housing with either ball or needle bearings. The rotor shown in Fig. 6-24 has fourteen poles, or seven pairs and, because of their intermeshing

Figure 6-22.
DC alternator stator.

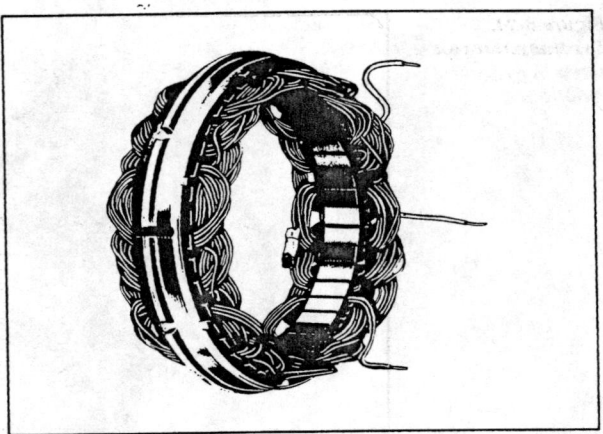

configuration, the poles alternate, north and south. DC from the voltage regulator energizes the drum-type coil in the core of this rotor. The magnetic field produces alternating north and south poles in the soft iron, interlacing fingers formed by the two end-pieces.

The rectifier of an alternator is made up of six silicon diodes, three mounted in the end frame, and the other three pressed into an insulated heat sink. Fig. 6-25 is a schematic of an aircraft alternator showing the three-phase Y-connected

Figure 6-23.
The three-phase AC output of the alternator stator is rectified by a six-diode, full-wave rectifier.

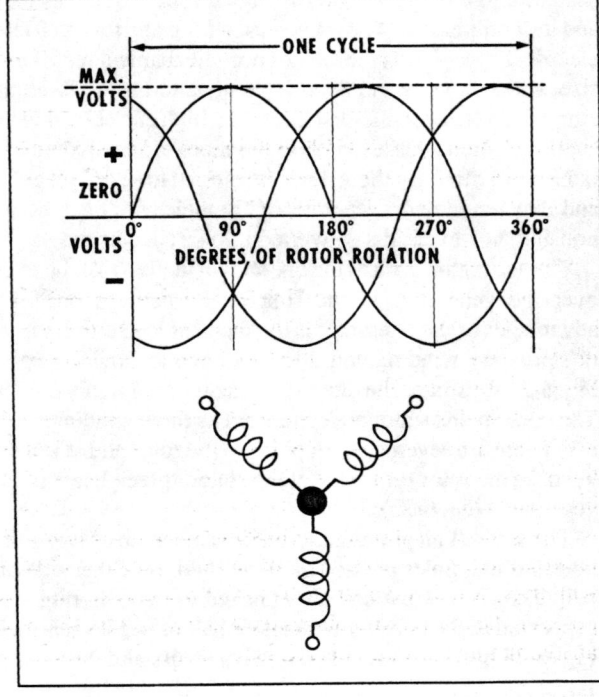

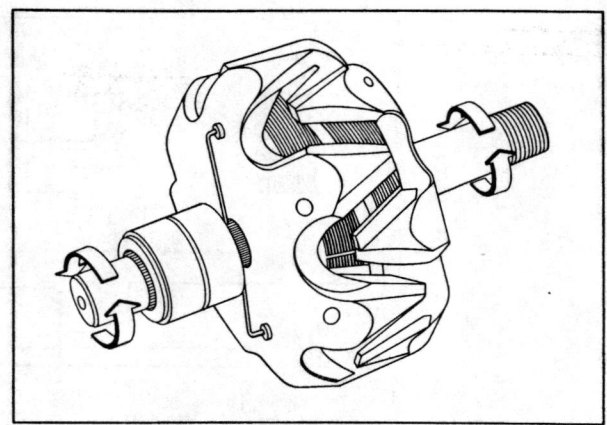

Figure 6-24.
DC alternator rotor.

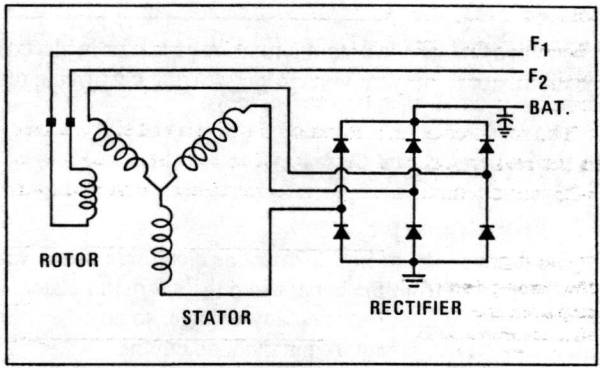

Figure 6-25.
DC alternator circuit.

F₁
F₂
BAT.

ROTOR

STATOR

RECTIFIER

stator, the single-phase rotor, and the six rectifier diodes. A capacitor is often placed at the output of the rectifier to protect the diodes from voltage surges caused by their blocking the current.

2. Brushless Alternators

High-output DC alternators may be of the brushless type. That is, there are actually two alternators in the same housing. One alternator produces DC for the load, and the other is controlled by the voltage regulator to provide excitation voltage for the output.

In Fig. 6-26 we have a schematic of a typical brushless alternator. The exciter is stationary and receives field DC from the voltage regulator. If output is low, the regulator will supply a larger amount of current than it does when output is high. The magnetic field of the exciter stator is cut by the exciter rotor, and the resulting three-phase AC is rectified by the three diodes which act as a half-wave rectifier. The rectified output of the exciter rotor flows into the output field windings, which are also a part of the rotor. This field creates north and south poles on the rotor frame, and as it rotates within the three-phase stator, generates an alternating current. This AC is fed into a six-diode full-wave rectifier, where

65

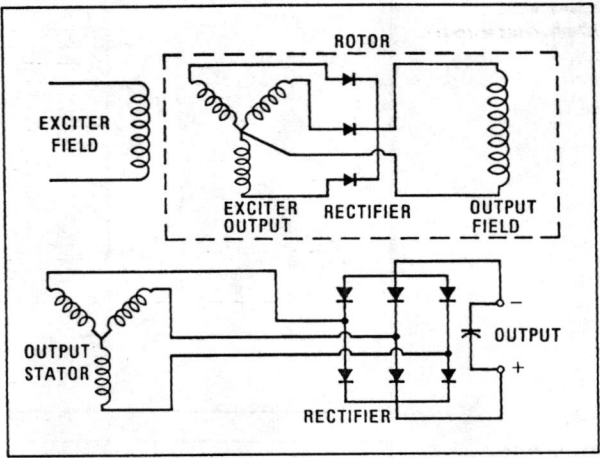

Figure 6-26.
Brushless DC
alternator circuit.

it becomes DC for the output circuit. A capacitor is placed across the output of the alternator to prevent voltage surges from destroying the rectifying diodes.

D. Alternator Controls

1. Vibrator-type

The diodes seen in Fig. 6-25 act as electronic check valves and prevent the flow of current from the battery into the alternator stator windings, even when the alternator is not producing any voltage, so no other form of reverse-current protection device is required in the load circuit.

The field excitation of an alternator is taken from *regulated* voltage rather than directly from the source, so the maximum amount of current the alternator can produce is automatically limited, and there is no need for a current limiter as we used with a generator. Two controls are needed however—one, a voltage regulator to control the amount of current flowing in the rotor, or field, and the other, a field relay to open the field circuit when the generator is not operating. This prevents the battery discharging through the field.

Fig. 6-27 is a schematic of a vibrator-type control for an alternator. When the alternator master switch is closed, the field relay is energized, closed, and allows current to flow from the battery, flowing through the field relay contacts, the voltage regulator lower and center contacts, and into the alternator field. The alternator can now produce electricity, and when the voltage rises to the regulated value, the voltage regulator coil opens the contacts. The field current must now flow through the resistor R_F to energize the field.

As long as the alternator is operating in the lower end of the regulated voltage range, the points are vibrating between the lower and center contacts, allowing the current to go directly to the field or to pass through the resistor. As the alternator speed increases, the points are pulled down so that as they vibrate, they alternate between placing the resistor in the field and grounding the field.

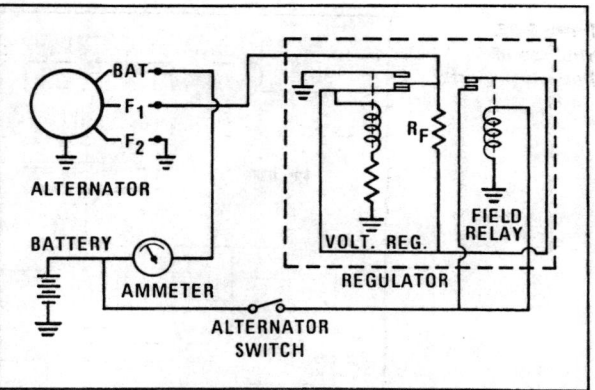

Figure 6-27.
Vibrator-type voltage regulator and field relay for a DC alternator.

With both ends of the field grounded, there is no current in the field and voltage drops off immediately.

2. Transistorized Voltage Regulator

Vibrating points are a source of mechanical problems, as well as arcing which can cause radio interference. The advent of the transistor and its use as a switching device has made it possible to control voltage of alternators without the use of the vibrating points. A transistor can be used as a switch and accomplish the same thing as the points. Let's review the transistor to see just how it works. (For a more complete treatment of the operation of semiconductor devices in general and transistors in particular, we refer you to the IAP training manual *Basic Electronics and Radio Installation.*)

If we have an insulating material, such as silicon, that will not allow any electron flow through it, we can place it in a circuit with a battery and an ammeter (Fig. 6-28A) and there will be no flow regardless of the polarity of the battery. Now, if we introduce into our silicon a few parts per million of an element that has an excess of electrons, we will have what is called an N-type silicon, which will behave differently from pure silicon. When a battery is connected across the N-type silicon, all of these extra electrons will be attracted by the positive charge of the battery and will all move to the side near the positive terminal.

If the silicon is doped with a material having a *deficiency* of electrons, there will be areas where there is need for electrons and it may be said that "holes" exist. This doped material is called P-type because there is a deficiency of electrons (negative charge). If a piece of P-type silicon is placed across a battery, all of the holes will be attracted to the side of the material nearest the negative terminal, but there will be no flow in the circuit (Fig. 6-28C).

Interesting things begin to happen, though, when we join a piece of N-material with a piece of P-material. And we can now see why this is called a semiconductor. If the two pieces are fused together and connected to a battery with the N-material to the positive terminal and the P-material to the negative, all of the holes will be drawn toward the negative side of the material and all of the electrons to the positive, and there will be no flow. The diode, as this device

67

**Figure 6-28.
Principle of
solid-state diode
action.**

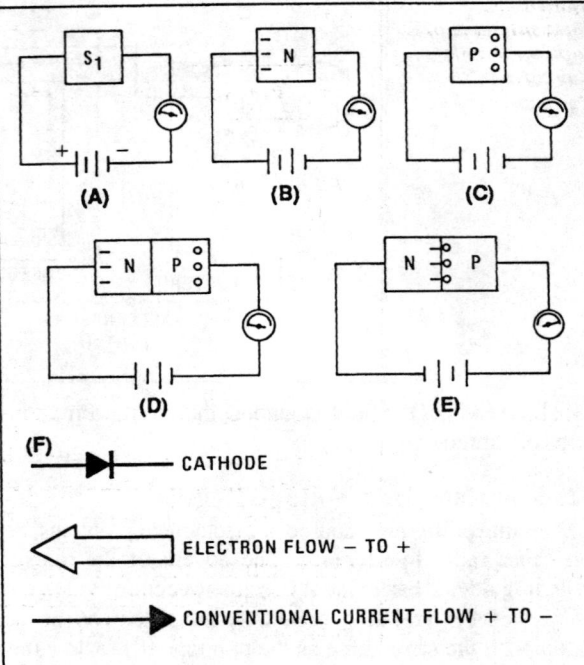

(A)   (B)   (C)

(D)   (E)

(F) ►▌──── CATHODE

⟸ ELECTRON FLOW − TO +

⟹ CONVENTIONAL CURRENT FLOW + TO −

(A) PURE SILICON WILL NOT PASS ANY ELECTRON FLOW.
(B) SILICON DOPED TO PRODUCE AN N-TYPE MATERIAL
     WILL NOT ALLOW ELECTRON FLOW.
(C) SILICON DOPED TO PRODUCE A P-TYPE MATERIAL
     WILL NOT ALLOW ELECTRON FLOW.
(D) N- AND P-TYPE MATERIALS JOINED IN A JUNCTION
     WILL NOT ALLOW ELECTRON FLOW WHEN THEY
     ARE REVERSE-BIASED.
(E) N- AND P-TYPE MATERIALS JOINED IN A JUNCTION
     WILL ALLOW ELECTRON FLOW WHEN THEY ARE *FOR-
     WARD*-BIASED.
(F) ELECTRON FLOW IS FROM NEGATIVE TO POSITIVE,
     AND IS AGAINST THE ARROW OF THE SEMICONDUCTOR
     SYMBOL; BUT CONVENTIONAL CURRENT, POSITIVE TO
     NEGATIVE, IS IN THE DIRECTION OF THE ARROW IN THE
     DIODE SYMBOL.

is called, is reverse-biased. If the battery is turned around as we see in Fig. 6-28E, both the holes and the electrons will be forced to the junction and current now flows. The diode is said to be forward-biased.

Semiconductor diodes are used as check valves for electron flow, and an elaboration of the diode called a transistor can be used as a control valve. In Fig.

Figure 6-29.
Basic principle of the transistorized voltage regulator.

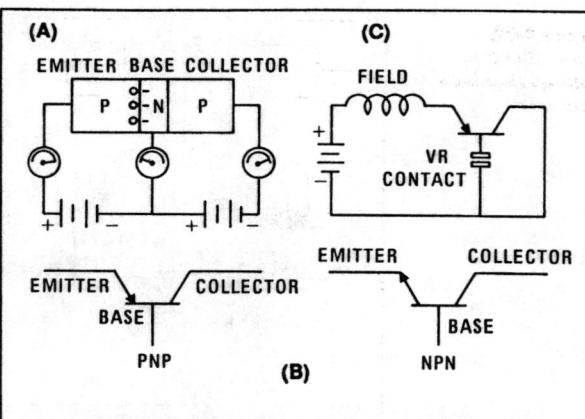

(A) A TRANSISTOR IS MADE UP OF A SANDWICH OF P-AND N-TYPE SILICON. EMITTER-COLLECTOR CURRENT FLOWS ONLY WHEN THE BASE CIRCUIT IS COMPLETE.

(B) NPN AND PNP TRANSISTORS DIFFER ONLY IN THE POLARITY OF THE VOLTAGE.

(C) THE ALTERNATOR FIELD IS IN THE EMITTER-COLLEC-TOR CIRCUIT, AND THE POINTS CONTROL ONLY THE EXTREMELY SMALL BASE CURRENT.

6-29A we see the way a transistor works. Let's assume that it is made up of a sandwich of N-type silicon between two pieces of P-type, with the N layer *extremely* thin. If we place a couple of batteries in the circuit such as we see here, the emitter-base junction will be forward-biased (+ to the P material and – to the N) and there will be some current flow between the emitter and base. The positive voltage of the battery will drive all of the holes in the P-type emitter to the junction where they combine with electrons from the N-type base.

Now we have said that the base was extremely thin, and the negative charge on the base has attracted more holes to its edge than the thin base can fill with electrons. But the collector is connected to the negative terminal of the battery, and it can supply all of the electrons needed. What really happens is, as long as there is a flow of base current which is extremely small, there will be a large flow of emitter-collector current. Any time the base current is stopped, no current can flow between the emitter and collector.

The symbols used for transistors are shown in Fig. 6-29B. The PNP and the NPN transistors are similar in operation, except for the polarity of the battery. It must be remembered that the flow of electrons is *opposite* the arrows, but conventional (+ to –) current follows them.

A practical use of the transistor is the control of alternator field current without using contacts to carry any heavy current. In Fig. 6-29C we have the basic circuit of a transistorized voltage regulator. The contact points in the base circuit are controlled by the alternator output voltage. When it reaches the regulated value,

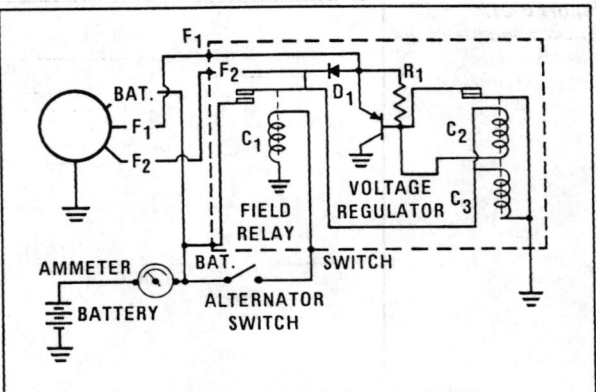

Figure 6-30.
Transistorized voltage regulator and field relay.

the contacts are pulled open and base current stops. And with no base current, there can be no field current, so the voltage drops until the contacts close, allowing it to build back up. There is no sparking with this type of regulator, as the contact points carry only an extremely small amount of current.

A practical circuit of a transistorized voltage regulator is seen in Fig. 6-30. When the alternator switch is closed, the field relay C_1 is energized, closing its contacts, and allowing current to flow into the field through F_2 and out F_1 to ground through the transistor. The transistor is conducting between its emitter and collector, because the voltage regulator contacts are closed and emitter-base current is flowing to ground through them. The output voltage will rise until the combined pull of C_2 and C_3 becomes strong enough to open the voltage regulator points.

Now, no base current can flow and the transistor stops conducting, so there is no more field current. While the points are open, current will still flow through C_3 and its magnetic field will still be active; but the circuit through C_2 is broken and its field instantly collapses, so the points will close rapidly. When the points close, the total pull will increase rapidly because of the two coils. This second winding is appropriately called an accelerator winding. The diode D_1 is used to prevent voltage surge, which occurs when current abruptly stops flowing through the field winding, from damaging the transistor. When the surge occurs, it causes the diode to break down and conduct through to the opposite side of the field until the voltage drops to a value that will not damage the transistor.

Resistor R_1 is in the circuit to help keep the transistor shut off when the contact points are open.

3. Transistor Voltage Regulator

While the transistorized voltage regulator uses a transistor to actually control the flow of field current, an electromagnetic coil is used to sense voltage. *Transistor* regulators, on the other hand, are fully solid-state; that is, there are no moving parts in the regulator itself. Voltage is sensed by a special kind of device called a zener diode.

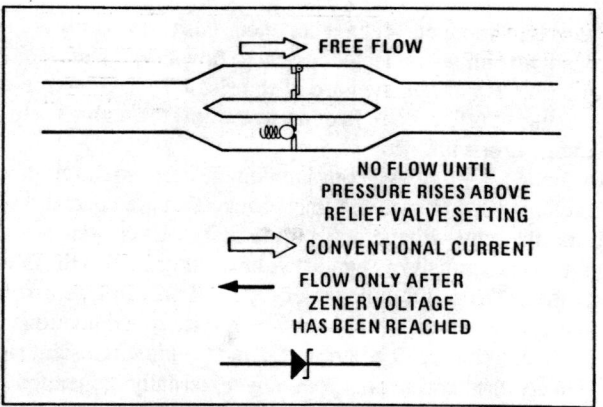

Figure 6-31.
Principle of a zener
diode.

FREE FLOW

NO FLOW UNTIL
PRESSURE RISES ABOVE
RELIEF VALVE SETTING

CONVENTIONAL CURRENT

FLOW ONLY AFTER
ZENER VOLTAGE
HAS BEEN REACHED

Fig. 6-31 explains the action of a zener diode, which acts as though it had a check valve in parallel with a relief valve. The combination allows free flow in one direction but prevents flow in the opposite direction, until the pressure rises above the value maintained by the relief valve. The zener allows free electron flow in one direction, but resists reverse flow until the voltage rises to its breakdown, or zener, value.

In Fig. 6-32 we have the complete circuit of a basic transistor voltage regulator. The output of the alternator is across the voltage divider made up of resistors R_1, R_2, and R_3. The zener diode D_1 senses the voltage drop across R_1 and part of R_2. When the voltage across it is low, there will be no current flow through the base of transistor T_1, called the driver transistor. And with no base current, there will be no emitter-collector current and no voltage drop across R_5. Base current can flow through transistor T_2 and it will conduct. T_2, you will notice, is the output transistor and is in series with the alternator field.

With the field receiving its full field current, output voltage will rise, and at the point of regulation, there will be enough voltage across the zener diode to

Figure 6-32.
Transistor voltage
regulator.

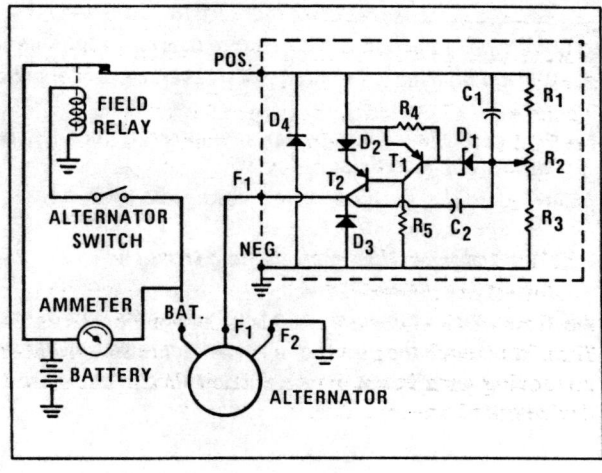

71

cause it to conduct. When it conducts, base current flows in transistor T_1, which causes an emitter-collector current to flow through resistor R_5. The voltage built up across R_5 brings the base of transistor T_2 to the same voltage as its emitter and shuts it off, so no field current can flow through T_2, and the alternator output voltage drops off.

Diode D_2 provides a constant voltage drop, so the emitter of transistor T_2 will be enough lower than the line voltage that the current through resistor R_5 can bring the base voltage of T_2 up to that of its emitter and shut it off. Diode D_3 protects the transistors against voltage surges when the field current is suddenly stopped. The rapid collapse of the field will induce a voltage high enough to damage the transistors, but before it rises to a dangerous value, D_3 will break down and conduct it to ground. Diode D_4 is a transient suppression diode that protects the transistors from any externally generated voltage surges, and capacitors C_1 and C_2 smooth out pulsations and cause the regulator to operate more smoothly.

QUESTIONS:

1. *If the fingers of the left hand encircle a conductor in the direction of the lines of magnetic flux, in what direction is the electron flow?*

2. *If the ends of an armature coil are attached to slip rings, which will be produced, AC or DC?*

3. *What three things determine the amount of current produced in an electro-magnetic generator?*

4. *What determines the distance the neutral plane of a generator moves away from the brushes?*

5. *Name three methods of minimizing brush arcing from armature reaction.*

6. *How is the voltage output of a generator controlled?*

7. *What makes a series-wound generator impractical for use in an aircraft power system?*

8. *Which field has the greater effect in a compound generator if the output voltage increases with an increase in load current?*

9. *What type of current limiter is used in large, high-output generators?*

10. *Why do alternators not require a reverse-current cut-out relay, as genera-tors do?*

11. *Why must generators which concurrently feed the bus of a multi-engine aircraft be paralleled?*

12. *What is the purpose of the accelerator winding on some vibrator-type voltage regulators?*

13. *What is done to the generator field circuit when the points of a vibrator-type voltage regulator open?*

14. *What causes the points of a reverse-current cut-out relay to close?*

15. *What causes the points of a reverse-current cut-out relay to open?*

16. *How does a vibrator-type current limiter decrease the generator current output?*

17. What is the difference between the Delco-Remy "A" circuit and "B" circuit generators?

18. What is the purpose of the upper set of points on some double-contact vibrator-type voltage regulators?

19. In a three-unit generator control system, which two units have heavy windings?

20. What two forces act in a carbon-pile voltage regulator to vary the force on the carbon pile?

21. What happens to the resistance of the carbon pile as generator output voltage increases?

22. In a twin-engine aircraft using carbon-pile voltage regulators, what is used to hold the output of the two generators equal?

23. What causes a differential-voltage reverse-current relay to take the generator off of the line?

24. What type of rectifier is used with a DC alternator?

25. Why does a DC alternator not require heavy brushes for the load current?

26. Why does a DC alternator normally put out its rated current at lower engine RPM than a DC generator?

27. How many diodes are used to form the rectifier in a DC alternator?

28. Why does a brushless alternator not require brushes for its operation?

29. Why does a DC alternator normally require no form of current limiter?

30. What is done in the field circuit of a DC alternator when the upper contact closes on a two-contact vibrator-type voltage regulator?

31. What is the function of the transistor in a transistorized voltage regulator?

32. If a semiconductor diode is installed in such a way that the positive terminal of the battery goes to the P-material and the negative terminal to the N-material, will it conduct?

33. What turns the transistor on and off in a transistorized voltage regulator?

34. What is different about the circuits which use PNP transistors and those which use NPN transistors?

35. How can a diode placed across the field windings in an alternator protect the transistor in a transistorized voltage regulator?

36. What is used to sense the voltage for controlling a transistor voltage regulator?

37. What is the function of the driver transistor in a transistor voltage regulator such as the one shown in Fig. 6-32?

Chapter VII
DC Generators And Controls—Overhaul And Servicing

A. Generator

An aircraft generator, as any aircraft component, must be maintained and overhauled in strict accordance with the manufacturer's recommendations. There is usually no definite time between overhaul stated for generators or alternators, but the normally accepted rule is for all of the accessories to be overhauled any time the engine is overhauled. The overhaul procedure will vary with the individual generator, but a few items are general enough to be mentioned here.

After the generator is removed from the airplane and taken into the shop, it is disassembled and cleaned. No part of a generator should be immersed in a cleaning vat, as the degreasers are likely to damage the wiring or the insulation; but grease and residue from the brushes may be removed with a rag or a brush, damp with an approved solvent, and then dried with compressed air.

Check the bearings and replace any that are rough or specified by the manufacturer to be replaced at overhaul. The overhaul manual will specify any type of bearing which can be cleaned and lubricated.

1. Fields

Field coils are not to be removed from the frame unless they must be replaced. Check them after they have been cleaned by connecting them in series with a heavy-duty rheostat, or carbon-pile resistor, across a battery (Fig. 7-1A). Adjust the resistor until proper voltage is applied across the coils, and if the current is not within the value specified by the manufacturer's overhaul manual, the field coil should be replaced. Inspect all of the wiring for indication of worn insulation and test for shorts between the field and the frame with a test lamp (Fig. 7-1B). Note: The test lamp should *not* light up.

2. Armature

Inspect the armature for any indication of a worn or bent shaft, separated laminations, bare wires, or damaged insulation. Check for shorts between the commutator and the shaft by using the test lamp. Place one test lead on the shaft and run the other around all of the segments of the commutator. The lamp should not light up. The commutator is made up of copper segments to which the ends

Figure 7-1.
Generator field tests.

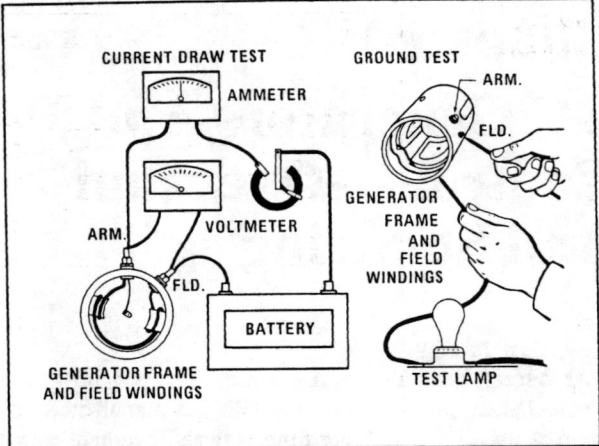

of the coils are soldered. Each segment is insulated from the shaft, the armature core, and from each other (Fig. 7-2A).

If the commutator is rough, burned, or out of round, or if the mica separators stick up above the copper, it should be turned. Chuck the armature in a lathe, and take very light cuts across the copper until it is clean and true. Undercut the mica with a blade about 0.025 inch thick, to a depth of about 1/32 inch and remove all of the burrs with fine sandpaper, *not* emery.

After the armature has been turned and visually inspected, it should be tested on a growler (Fig. 7-2C). A growler consists of a heavy horseshoe-shaped frame, made up of laminated steel with a coil of wire in its center, excited with 60-hertz AC. The armature to be tested is laid across the open ends of the frame, and the growler winding then acts as the primary of a transformer and the windings of the armature as the secondary. A strip of steel, such as a hacksaw blade, is held above the armature and the armature is rotated.

The magnetic effect on the hacksaw blade gives a pretty good indication of the uniformity of the windings. If any of the windings are shorted, their magnetic effect will be greater than that of the unshorted coils, producing a much stronger attraction for the hacksaw blade. Internal shorts will cause the metal strip to vibrate violently.

3. Brushes

When the fields, the armature, and the bearings are all determined to be in good condition, the generator may be reassembled. New brushes should be installed and seated by holding a strip of fine sandpaper between the commutator and the brush, with the grit toward the brush. Pulling the strip back and forth will quickly wear the brush to the proper contour of the commutator. Be sure after seating the brushes to blow every trace of sanding dust from the housing. Check the brush spring tension by using a spring scale and measuring the pull required to just raise the arm from the brush.

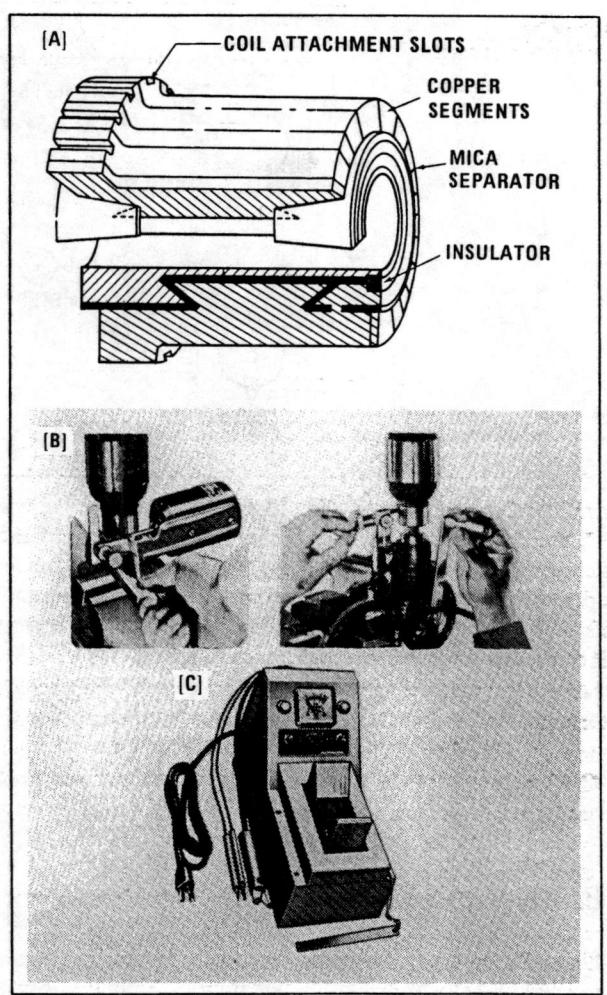

Figure 7-2. DC generator armature.

[A]
COIL ATTACHMENT SLOTS
COPPER SEGMENTS
MICA SEPARATOR
INSULATOR

[B]

[C]

4. Testing

After the generator is completely assembled, it should be tested as a motor. If the field is connected to the insulated brush inside the generator, ground the external field terminal (Fig. 7-3A); but if the field is grounded inside the generator, connect a short jumper between the A and the F terminals. Connect a battery of the proper voltage with an ammeter and a heavy carbon-pile resistor in series, and place a voltmeter across the generator output. Decrease the resistance of the carbon pile until the voltage specified in the manufacturer's test procedure is across the generator. The current should be within the manufacturer's tolerance, and the rotor should turn freely. If the field wire inside the generator attaches to the grounded brush (Fig. 7-3B), a jumper must be placed between the armature and the field terminals to cause the generator to motor properly.

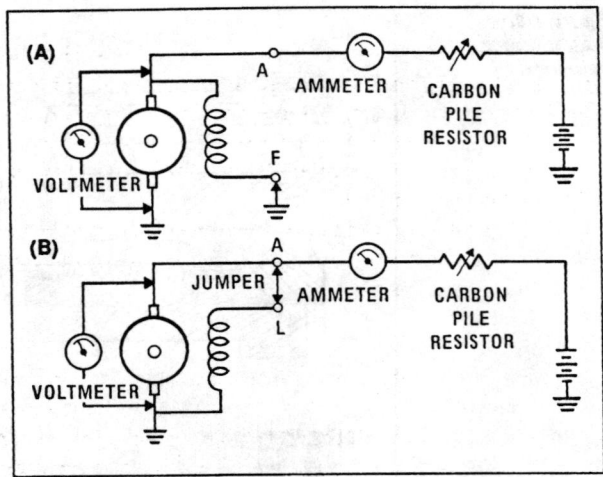

Figure 7-3.
Motoring test for a DC generator.

When the generator is being installed, and before the wires are connected, it is possible to lose all of the residual magnetism in the field frame. So after the generator is installed and all the wires connected, but before the engine is operated, the generator should be polarized—or, as it is sometimes called, have its field flashed. If the generator has a regulator in the ground end of the field, it is polarized by *momentarily* touching a wire between the positive polarity of the battery and the A terminal of the generator. This will send a surge of current through the field in the proper direction and magnetize the cores of the field coils. If the generator has a field winding connected to the ground brush, disconnect the field lead from the regulator and *momentarily* touch it to the battery terminal.

B. Alternators

1. Stator

The alternator is becoming much more popular now than it has been in the past because of its charging capability at low speed and its relatively trouble-free life. Like the generator, it should be overhauled at the time of engine overhaul, and, as with every aircraft component, the manufacturer's recommendations are the final authority for this work.

After the alternator is removed from the aircraft, it is disassembled and cleaned, not by soaking it in a decarbonizer, but by wiping it with a rag damp with the proper solvent. Stator leads are disconnected from the diodes and checked with a 110-volt continuity light or an ohmmeter. There should be continuity between any two of the three leads, but there should be no continuity to ground. If there is no indication of worn insulation, breaks, or damage to the stator housing, and no separated laminations, the stator can be considered satisfactory for reassembly.

2. Rotor

The rotor of an alternator is the field, and DC flows in its coil. Check the rotor for physical conditions: the shaft should be straight and free from any scratches or indications of damage. If there are any damaged threads, they should be dressed with a triangular file. The slip rings should be free from any scratches, and they should be tight on their insulator ring; and the wires feeding into the coil should not be damaged.

Check the resistance of the coil with an ohmmeter to see that it is the value indicated on the alternator test sheets. Use a 110-volt test light to test for shorts between the slip rings and the shaft; the lamp should not light up.

3. Diodes

A semiconductor diode is an electronic check valve, and to check it, use an ohmmeter set on its lowest (R x 1) range. Determine that the ohmmeter does not have more than 1½ volts between its test leads by placing them across a low range DC voltmeter. If the ohmmeter is properly set for checking alternator diodes, measure the resistance across the diode first with the red lead to the diode body; then reverse the leads and again measure the resistance. This procedure first forward-biases the diode and measures the current flow, and then reverse-biases the diode and measures the flow. In one position the ohmmeter should indicate a relatively high resistance, and in the other the resistance indication should be low. *Do not use the 110-volt test lamp to check a diode, as it will surely damage it.*

4. Brushes

Alternators do not have commutators, and the brushes do not carry large amounts of current; but even a small brush, carrying only field current into the smooth slip rings can wear, and, at overhaul, the brushes should be checked to determine that they are within the length specified by the manufacturer, and that they operate smoothly in the brush housing.

5. Testing

After the alternator has been reassembled, all of the wires properly connected, and the rotor checked for smooth rotation, it is placed on a test stand; here a battery in parallel with a variable-load resistor is placed between the alternator's "Batt." terminal and ground. Ground one of the field terminals and connect the other with a jumper to the Batt. terminal. Operate the alternator at the speed recommended in the test procedure, and adjust the load resistor to determine that the alternator will put out the rated voltage and current.

C. Generator And Alternator Controls

Voltage regulators and current limiters, while seeming to be simple in their construction and operation, require knowledge and care in their adjustment. No attempt should be made to adjust any generator or alternator control without the appropriate manufacturer's manuals and the recommended test equipment.

Many generator controls are temperature compensated; this means that bi-metallic compensators are incorporated in their manufacture that increase the output voltage when the regulator is cold and lower it when the regulator is hot.

When adjusting any generator control, mount it on the test stand in the same position in which it will be when installed in the airplane, and operate it to the point where the temperature is within the recommended limits before it is adjusted. The final settings are checked with the cover in place.

1. Vibrator-type Voltage Regulator

Before any electrical adjustments to this type of regulator are made, the air gap between the coil winding core and the armature should be set to that specified by the manufacturer. This measurement is made with the armature pushed down until the points are *just* making contact. After the air gap has been set, the regulator is placed in the circuit of the proper test equipment and the generator operated at specified speed. Voltage is adjusted by varying the tension on the voltage regulator contact spring.

2. Vibrator-type Current Regulator

Like the voltage regulator, the current regulator or limiter has two adjustments: the air gap setting and the tension on the contact spring. Place just enough pressure on the contact arm with your fingers that the points are *just* making contact, then measure the clearance between the coil core and the armature. If this is not within the tolerance specified, adjust it according to the manufacturer's recommendation.

With the air gap properly adjusted, place the regulator on the test stand with an accurate ammeter and a load in series with the battery lead. Run the generator until the regulator is up to the specified temperature, with a jumper across the voltage regulator points so that they will not open the field circuit; then increase the load to that specified for the particular test. Adjust the tension on the contact spring until the regulator prevents the generator putting out current in excess of that specified by the manufacturer.

3. Reverse-current Cut-out Relay

Reverse-current cut-out relays require three adjustments: the air gap, the amount the points open, and the closing voltage.

To adjust the air gap, the armature is pressed down with the fingers until the points *just* close, and the clearance between the armature and the coil core is measured. This should be within the tolerance specified by the manufacturer.

A flat spring holds the points open, and there is a stop that limits their maximum opening. This stop is bent to provide proper clearance between the points.

For the final adjustment of the reverse-current cut-out, mount the regulator on the appropriate test stand and increase the voltage of the generator until the reverse-current cut-out points close. This voltage is measured with an accurate voltmeter placed between the "Gen." terminal of the regulator and ground. If the opening voltage is not within the tolerance specified by the manufacturer, it is adjusted by varying the spring tension on the points.

4. Transistorized Voltage Regulators

The voltage coil on a transistorized voltage regulator is adjusted in the same way as that of the vibrator-type regulator. The air gap is adjusted first, and with the alternator and regulator on the test stand, the tension of the contact spring is adjusted to bring voltage into the allowable range.

5. Transistor Voltage Regulator

Many transistor voltage regulators are sealed units with no adjustment available. If all of the appropriate tests show that voltage is not within the proper operating range because of the regulator, the regulator is replaced. Other regulators have a variable resistor that may be adjusted from the outside of the case. The range of voltage adjustment is limited, but it is enough for paralleling the generators in a twin-engine installation.

QUESTIONS:

1. *How can you determine whether or not the field coils in a generator are internally shorted?*

2. *What would be indicated when testing an armature on a growler if the hacksaw blade vibrated excessively at one point?*

3. *What should be done to the mica insulation between the commutator segments when the commutator has been turned on a lathe?*

4. *Why is it important to seat the brushes when new ones are installed in a generator?*

5. *What is the purpose of flashing the field of a newly overhauled generator?*

6. *Should the engine be running or not running when the generator field is flashed?*

7. *How can you determine the condition of an alternator diode by using an ohmmeter?*

8. *Should the field of a newly overhauled alternator be flashed?*

9. *What two adjustments are made on a vibrator-type voltage regulator?*

10. *What three adjustments are made on a reverse-current cut-out relay?*

Chapter VIII
AC Alternators And
Their Controls

A. Three-phase Alternators

Large aircraft which have electrical power requirements beyond that which can be provided with direct current systems use alternating current systems. Electric power, as you remember, is the product of the voltage and the current, and since the voltage must be held low in a direct-current system, the current must, of necessity, be high, and this requires large and heavy wiring. Alternating current systems use three-phase alternators which produce 120 volts across each phase and 208 volts across any two of the phases. Fig. 8-1 shows the schematic of a typical brushless alternator with a single-phase excitation winding for the field and a three-phase Y-connected output.

Commercial alternating current in the United States has a frequency of 60 hertz, but in order to use smaller and lighter weight components, aircraft AC has been standardized at 400 hertz. In order to achieve a constant frequency from an alternator, which is driven by a variable speed engine, a constant speed drive unit (CSD) is installed between the engine and the alternator.

A CSD consists of a piston-type hydraulic pump splined into an engine accessory drive. The output of the pump drives a piston-type hydraulic motor whose speed is sensed by a flyweight-type governor. This motor drives the alternator, and if the alternator begins to slow down, the governor will cause the pump to put out more fluid, bringing the motor back to its governed speed.

Figure 8-1.
Three-phase
brushless AC
alternator.

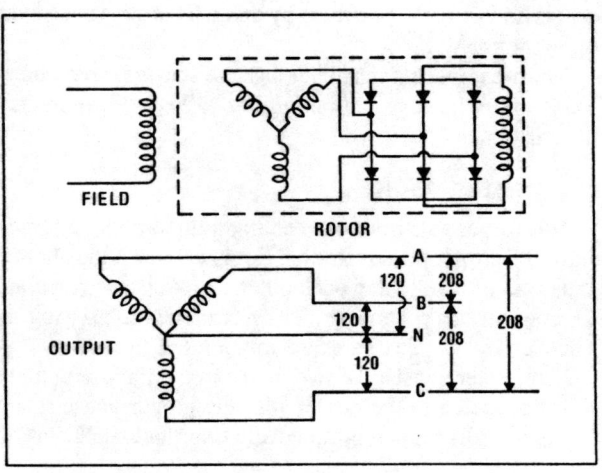

B. Alternator Controls

DC controls are primarily concerned with keeping voltage within rather close limits; limiting current to a value less than the rated output of the generator; and preventing a reverse flow of current through either the field or the armature windings. AC systems are more demanding in their controls, and must sense or control several parameters other than just voltage and current.

1. Underfrequency Sensor

Inductive reactance limits the flow of current in an AC system, and its value is determined by frequency. If AC frequency drops, there is less inductive reactance to oppose the flow of current, and components could be damaged. To prevent this excess current, a frequency sensor decreases excitation current to the alternator field any time output frequency drops below a specific value.

2. Line Voltage Sensing

A voltage sensor measures the voltage of each of the phases and decreases the output if any phase gets above the limit set.

3. Reactive Power Division

Real power in an AC circuit is the product of the voltage and that portion of the current which is in phase with the voltage. It is expressed in watts and is found by multiplying the voltage by the current, and this by the power factor. (Power = $E \times I \times$ Power Factor.) Reactive power, on the other hand, is the product of the voltage and the reactive current, some of which is not in phase with the voltage. Reactive power is expressed in VARs (volt-amperes, reactive) or KVARs (kilovolt-amperes, reactive.)

The amount of torque required to drive an alternator is determined by the true power it is producing; but the excitation required is determined by the reactive power. When alternators are operating in parallel and reactive loads are not balanced, it is possible for one alternator to draw some reactive current from others and actually be driven by them. This can cause overheating and result in alternator damage.

Multi-engine AC installations have sensors which measure relative reactive loads and adjust excitation current to keep these loads balanced among the alternators.

4. Phase Rotation

It is important, when connecting a three-phase alternator into the circuit, to have the proper phase rotation. If it is reversed, the alternators will oppose each other rather than aiding each other. Proper phase rotation is assured by making all connections according to the manufacturer's wiring diagram, and it may be checked by the use of a phase-rotation test light.

This tester consists of two lamps in a circuit with a phase-shift capacitor. If it is connected in the circuit with the proper phase rotation, the lamp marked "Bright" will burn bright and the one marked "Dim" will dim. If the phase

84

rotation is incorrect, the brilliance of the lamps will be *opposite* to the way they are marked.

QUESTIONS:

1. What is the frequency of most aircraft AC systems?
2. What is used in aircraft AC power systems to keep the frequency of the alternator constant as the speed of the engine varies?
3. What is the difference between real power and reactive power?

Chapter IX
Electrical Motors

A. DC Motors

An electrical motor is similar to a generator in that they are both energy exchange devices. The generator receives mechanical energy from the engine and converts it into electrical energy as a conductor is moved within a magnetic field. A motor, however, receives electrical energy, and by the action of two magnetic fields produces mechanical energy.

Figure 9-1A: When current flows in a conductor placed in a magnetic field, the combined forces of the two fields will cause the conductor to move.

Figure 9-1B: The reaction between the flux of the armature and that of the field causes the armature to rotate in the direction shown.

Figure 9-1C: The field and armature of a shunt-wound motor are in parallel.

In Fig. 9-1A, we see a current-carrying conductor in the magnetic field of a permanent magnet. If we think back to the left-hand rule regarding current flow in a conductor, we remember that if we will grasp a conductor by the left hand with the thumb pointing in the direction of electron flow, the fingers will encircle the conductor in the direction of the lines of flux. We see in Fig. 9-1 that a conductor carrying current away from the viewer will have lines of flux encircling it in a counterclockwise direction, and the lines from the permanent magnet will extend from the north pole to the south. The lines of flux will add, below the conductor, and subtract above it, with the result that the intensified field below the conductor will force it upward.

Figure 9-1.
Shunt-wound DC motor.

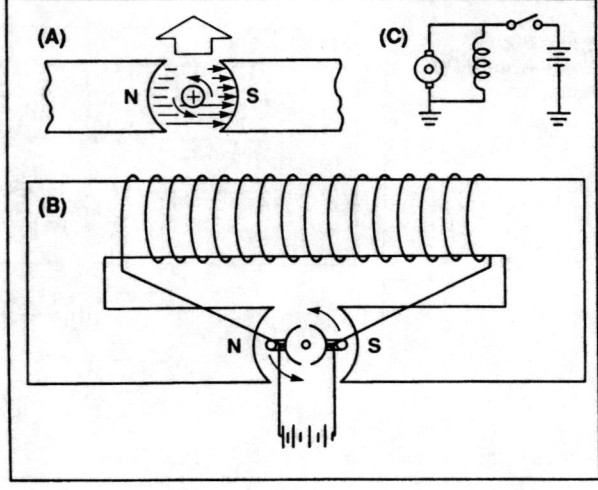

This result is also in keeping with the right-hand rule for motors which states: *"If the thumb, forefinger, and middle finger of the right hand are arranged so they are all perpendicular to each other, when the forefinger points in the direction of the flux and the middle finger in the direction of the electron flow in the conductor, the thumb will point in the direction of movement of the conductor."*

Small, low torque motors may be built using permanent magnetic fields, and these motors may be reversed by simply reversing the direction of the current in their armature. When more torque is required, motors are built with electromagnets for their field. Fig. 9-1 shows a DC shunt motor. The battery sends current into the armature through the brushes and commutator and produces magnetic fields around the armature windings. The field flux is provided by the winding around the frame, which is in parallel, or shunt, with the armature.

Shunt motors have the characteristic of relatively low starting torque but tend to operate at a fairly constant speed. The reason for this can be seen as we visualize the armature and field in parallel. Current flows in both the armature and field, but the armature resistance is so much lower than that of the field that the field gets relatively little current, and it does not produce much flux, so the starting torque is low. When the rotor starts to turn in the magnetic field, it acts not only as a motor, but also as a generator and produces a voltage.

This voltage, the counter electromotive force (CEMF), is opposite to the voltage that caused it, and so it decreases net voltage across the armature. Then, as armature net voltage decreases, so does armature current, and this allows field current to increase until the motor reaches a constant speed. If a load tends to slow the motor down, the decreased CEMF will allow more armature current to flow, increasing the torque, and bringing the motor back to speed.

Engine starters, landing gear, and flap motors all operate into a load at all times, and all require a high starting torque. For this reason they are series wound (Fig. 9-2).

Figure 9-2.
Series-wound DC
motor.

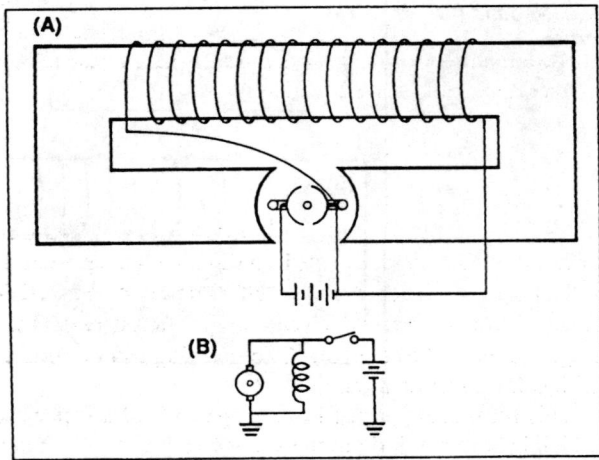

Figure 9-2A: The armature and field windings are in series in this motor.

Figure 9-2B: All of the current that flows in the armature also flows in the field winding and produces a high starting torque.

When the switch is closed, current flows through both the low resistance armature and the low resistance field coils. This high current produces two strong magnetic fields and provides sufficient torque to turn the engine, or raise or lower the landing gear. As armature speed begins to build up, it generates CEMF, and both current and torque decrease; but if the motor is loaded and slows down, CEMF decreases, allowing load current and torque to build up.

Series-wound motors must be operated into a load because of a characteristic which will cause their speed to increase until they can actually tear themselves apart. Let's assume a condition in which a series-wound motor is operated without a load connected. As armature speed increases, CEMF builds up, decreasing current through both the armature and the field. This decrease in field current drops the CEMF enough to increase net voltage across the armature, and it speeds up.

Compound-wound motors have both a series and a shunt field and provide sufficient starting torque for the load they are designed to carry; at the same time they have enough constant speed characteristics that they will not run away and tear themselves up.

B. AC Motors

1. Universal Motors

Electric drills and various other portable electric tools use universal motors, that is, a motor which will operate on either AC or DC. These motors are identified by their having an armature with a commutator and brushes and field windings similar to those used in DC motors. The reason this kind of motor can be operated on AC as well as DC is that when the polarity of AC voltage changes, as it does twice in each cycle, the polarities of *both* the armature and the field windings change at the same time, and the push is always in the same direction on the armature windings. If only the armature *or* the field polarity changed, the motor would reverse its direction.

2. Induction Motors

a. Three-phase

Induction motors are often called "squirrel cage" motors because of the type of rotor used. A stack of soft iron discs with slots around their circumference and a hole in their center is pressed onto the rotor shaft (Fig. 9-3). The slots are filled with copper bars which extend over the edges of the stack, and copper end rings are welded to the bars. The resulting rotor resembles a tread mill filled with soft iron lamination.

This rotor is supported in bearings and rides inside the three-phase stator, which is built similarly to the stator found in a DC alternator (Fig. 6-22). The

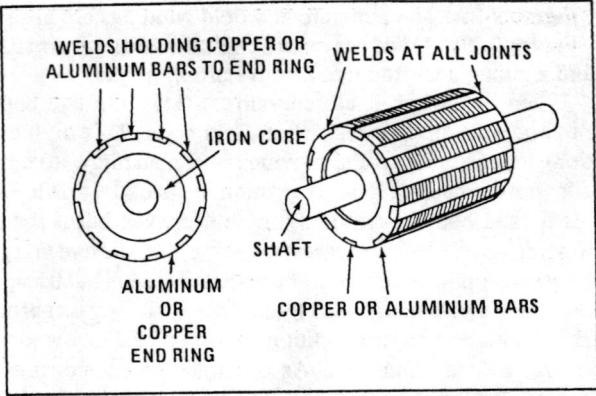

Figure 9-3.
The squirrel cage rotor of an AC induction motor.

WELDS HOLDING COPPER OR
ALUMINUM BARS TO END RING

WELDS AT ALL JOINTS

IRON CORE

SHAFT

ALUMINUM
OR
COPPER
END RING

COPPER OR ALUMINUM BARS

three-phase windings may be connected in either Y or Delta, depending on the design requirements of the motor.

When current begins to flow in one of the windings, a voltage is induced into the winding in the rotor. You will remember that these are large copper bars welded into copper end rings, and the resistance of this loop is just about as low as you can get it. There is no insulation between the copper and the iron laminations, but the laminations are somewhat insulated from each other, and there is little tendency for current to flow in the core. The voltage induced into the bars causes current to flow in them, and this in turn sets up a magnetic field which causes the rotor to turn in its bearings. As soon as the voltage drops in one phase winding, it is building up in the next, and the magnetic field of the stator rotates. The rotor spins as it attempts to lock its field with the rotating field of the rotor. The speed of the motor is fixed, depending on the frequency of the AC and the number of poles in the motor. When the motor is loaded, the rotor attempts to slow down, but instead of being able to, its slip, or the angular distance between the magnetic field of the rotor and that of the stator, increases, which increases the torque and brings the rotor back up to speed.

b. Single-phase

1) Capacitor-start Motor

Single-phase induction motors are built in much the same way as three-phase motors except that single-phase AC produces no rotating field. A single-phase motor will run equally well in either direction once it has been started, but it must have some provision for starting, and getting the rotor up to near the operating speed.

A single-phase, capacitor-start induction motor has two windings, one for starting, and one for running. The run winding is permanently connected across the AC input line (Fig. 9-4) but the start winding is attached to the power through a centrifugal switch. When the motor is not running, the switch is closed, but when the motor starts and comes up to about 75% of its operating speed, the switch opens and takes the start windings out of the circuit. The start winding is in series with a capacitor, which causes its current to lead the voltage and

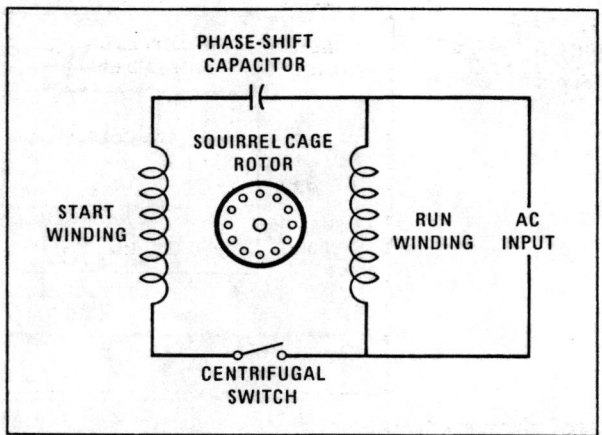

Figure 9-4.
A single-phase, capacitor-start AC induction motor.

provides a phase difference between the start and the run windings. The two sets of windings are wound in alternate slots in the motor's stator so that a rotating field is set up to get the rotor started. Once the rotor is spinning, it will continue to run without the need of the start winding.

2) Shaded-pole Motor

Low torque induction motors having a relatively constant speed may be made to operate on single-phase AC without the complication of the start windings or a centrifugal switch; these are called shaded-pole motors because of the construction of their stator.

Fig. 9-5 is a diagram of the most simple and one of the most commonly used shaded-pole motors. The rotor is similar to any other induction motor in that it has a laminated iron core with copper bars and end pieces. The stator for the motor is a horseshoe-shaped frame made up of laminated soft iron sheets, and around the frame is wound a coil of fine wire which is excited with alternating current. You will notice that around two edges of the frame, next to the armature, are slots in which there is a single wrap of heavy copper wire called a shading coil.

Let's see the way this motor operates. In Figure 9-5A, the low torque shaded-pole motor uses a squirrel cage rotor and a single-phase coil around the frame for excitation. Next, we see the flux lines without the shading coils (Fig. 9-5B, C). In one half-cycle of AC the lines of flux will flow as shown in B and will induce a current into the copper bars of the rotor which will magnetize its core as shown. In the next half-cycle (Fig. 9-C) the polarity will reverse and the lines of flux will flow as shown here. The current flowing in the rotor bars will also reverse. Now if the rotor could be spun so the polarity of its field would not reverse, it would lock in and run as any other single-phase induction motor.

We need however, some method of starting the rotor into motion, and this is the purpose of the shading coils. Look at Fig. 9-5D. During the rise in current in the half-cycle when the left-hand side is being made a south pole, the *changing* current cuts across the single turn of the shading coil and induces a current in

Figure 9-5.
Shaded-pole AC
induction motor.

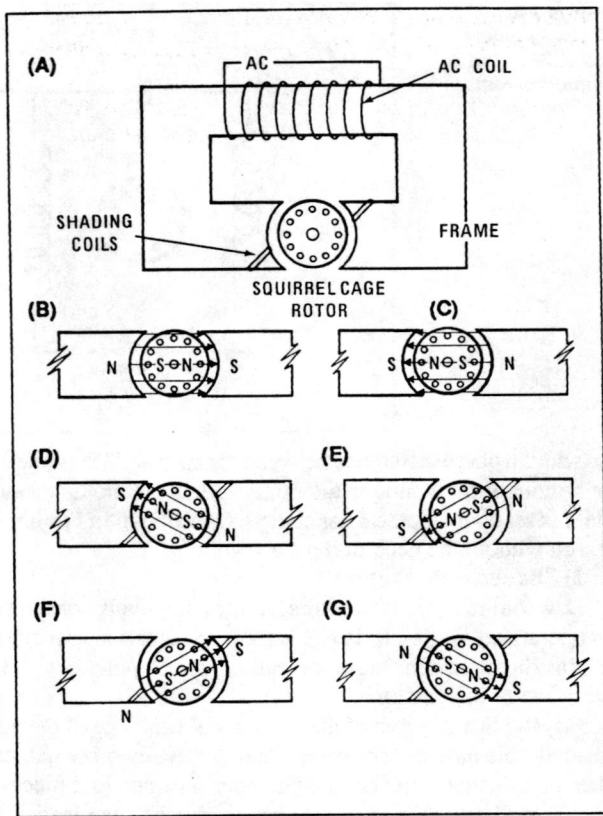

it, just as it induces a current into the rotor bars. The current flowing in the shading coil also has a magnetic field, and as we remember from Lenz's Law, the field from an induced current will be opposite to the field that caused it.

The result is that this field partially cancels the magnetism in the lower left side of the field opening. The concentration of the flux will flow diagonally across the armature. Now when the current reaches its peak and starts to drop off, the field of the shading coil tries to maintain the south polarity of this side, and the concentration of flux moves down and flows across from the sides with the shading coils (Fig. 9-5E); then, as it drops off again, the concentration is aligned with the shading coils. Looking at the rotor in Figs. 9-5D through G, we see that it follows the rotating field caused by the inductive action of the shading coils.

QUESTIONS:

1. *Which type of motor would have the greater amount of starting torque, a series-wound or a shunt-wound?*
2. *What will likely happen to a series motor if it is not operated into a load?*
3. *How can the direction of rotation of a DC motor be reversed?*

4. What is used for the rotor of an AC induction motor?
5. What is the purpose of the capacitor in a capacitor-start, single-phase AC induction motor?
6. How can a shading coil produce a rotating field in a single-phase AC induction motor?

Chapter X
Electrical System Installation

A. Power Requirements

Almost all aircraft currently produced in the United States are built under the provisions of FAR Part 23, which covers Normal, Utility, and Acrobatic category aircraft, or FAR Part 25, which covers Transport category aircraft. The electrical systems required for Transport category aircraft are so complex that we will make no attempt to cover their requirements here; but the type of aircraft normally used in general aviation meet the requirements of FAR Part 23, which we will discuss.

An analysis must be made of the electrical load to assure that the power source has adequate capacity to carry all of the load, and that all of the wiring, cabling, and circuit protection devices are adequate for the load. The installation must be made in such a way that the risk of electrical shock to the crew, passengers, or ground personnel is reduced to a minimum.

Electrical power sources of multi-engine aircraft must function either separately or in combination, and the failure of any component in one system must not impair the ability of any part of the other system to function properly.

There must be at least one generator to supply all of the electrical energy that is required for safe operation of the aircraft. This generator must be capable of continuously supplying its rated power, and it must be provided with adequate voltage control and provisions for preventing reverse current from draining the battery after shutdown. Suitable indicators must be provided to inform the pilot or crew of failure of the power source, and there must be a way for the pilot or crew to know the amount of power each source is producing.

Storage batteries must be installed in such a way that they will not overheat during a charge at maximum regulated voltage, and they must be so vented that any explosive or toxic gases will not cause a hazard, and no corrosive fluids can cause damage to any of the structure or adjacent essential equipment.

Fuses or circuit breakers must be incorporated in any circuit which could cause a hazard, except for the main circuit of the starter. No more than one essential circuit can be protected by a single circuit breaker or fuse, and if a circuit breaker is used, it must be of trip-free design, meaning that it cannot be closed into a fault, regardless of the position of its control. Automatic resetting circuit breakers are not permitted. If fuses are used, provisions must be made to carry one spare fuse of each rating, or 50% spares for each rating, whichever is greater.

A master switch is required that will disconnect the electrical power sources from the main bus, and the point of disconnection should be as near the source as practical.

The electrical wiring must be of sufficient size that it will carry the load without overheating from the current, nor should it have an excessive voltage drop.

All of the switches must carry the load for which they are intended, and all of them must be labeled so they are easily identified in flight.

These requirements for an electrical installation are merely excerpts from FAR Part 23, and before any alteration is made to an aircraft, the requirements should be studied in detail. Also to be studied are such methods as have been used and approved and are shown in Advisory Circular 43.13-1A and 2: Acceptable Methods, Techniques, and Practices for Aircraft Inspection and Repair, and for Aircraft Alteration.

B. Components

1. Wire

Aircraft electrical systems are wired with either MIL-W-5086 stranded copper wire or MIL-W-7072 stranded aluminum wire. Fig. 10-1 is a chart of both

COPPER WIRE SIZE SPECIFICATION MIL-W-5086	SINGLE WIRE IN FREE AIR MAXIMUM AMPERES	WIRE IN CONDUIT OR BUNDLED MAXIMUM AMPERES	MAXIMUM RESISTANCE OHMS/1000 FT. (20° C)	NOMINAL CONDUCTOR AREA CIRCULAR MILLS	FINISHED WIRE WEIGHT POUNDS PER 1,000 FEET
AN—20..................	11	7.5	10.25	1,119	5.6
AN—18..................	16	10	6.44	1,779	8.4
AN—16..................	22	13	4.76	2,409	10.8
AN—14..................	32	17	2.99	3,830	17.1
AN—12..................	41	23	1.88	6,088	25.0
AN—10..................	55	33	1.10	10,443	42.7
AN—8....................	73	46	.70	16,864	69.2
AN—6....................	101	60	.436	26,813	102.7
AN—4....................	135	80	.274	42,613	162.5
AN—2....................	181	100	.179	66,832	247.6
AN—1....................	211	125	.146	81,807	
AN—0....................	245	150	.114	104,118	382
AN—00..................	283	175	.090	133,665	482
AN—000................	328	200	.072	167,332	620
AN—0000..............	380	225	.057	211,954	770

ALUMINUM WIRE SIZE SPECIFICATION MIL-W-7072	SINGLE WIRE IN FREE AIR MAXIMUM AMPERES	WIRE IN CONDUIT OR BUNDLED MAXIMUM AMPERES	MAXIMUM RESISTANCE OHMS/1000 FT. (20° C)	NOMINAL CONDUCTOR AREA CIRCULAR MILLS	FINISHED WIRE WEIGHT POUNDS PER 1,000 FEET
AL—6....................	83	50	0.641	28,280	....
AL—4....................	108	66	.427	42,420	....
AL—2....................	152	90	.268	67,872	....
AL—0....................	202	123	.169	107,464	166
AL—00..................	235	145	.133	138,168	204
AL—000................	266	162	.109	168,872	250
AL—0000..............	303	190	.085	214,928	303

Figure 10-1. Copper and aluminum wire current-carrying capacities.

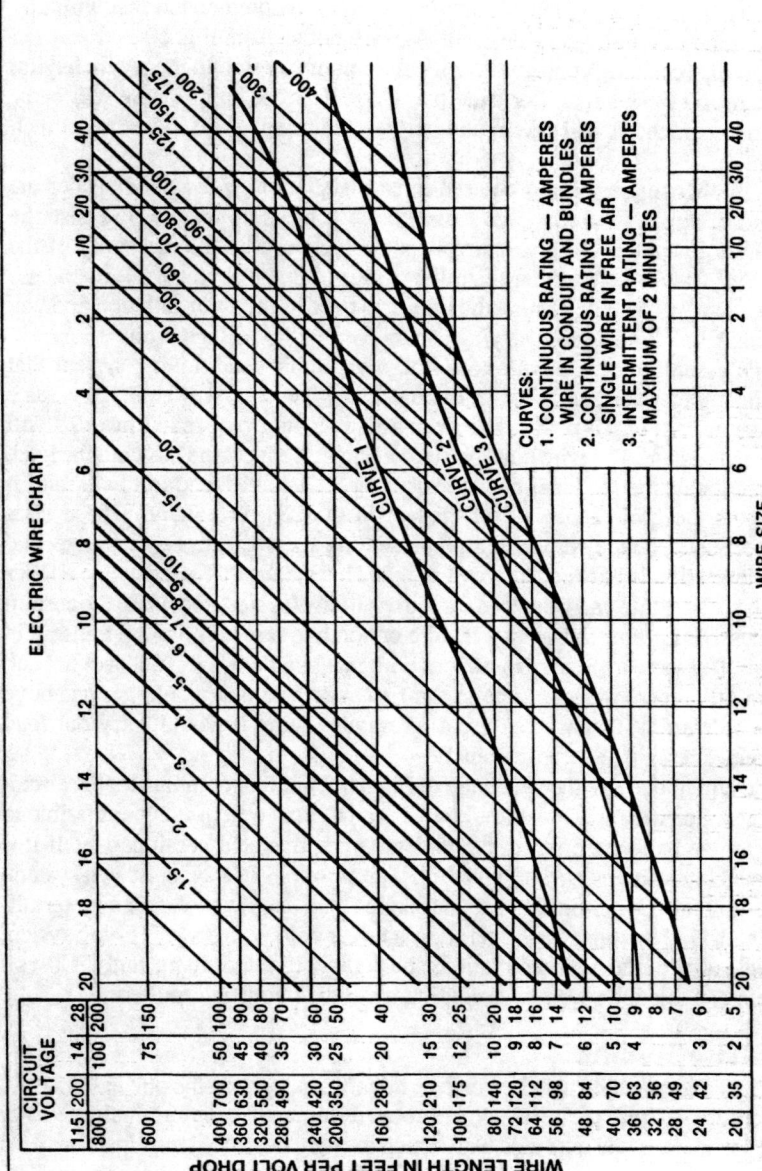

Figure 10-2 Electric wire chart.

97

copper and aluminum wire, giving its current-carrying capacity, both in free air, and in a bundle, its resistance, conductor area, and its weight. Because of the difficulty in terminating aluminum wire, it is not recommended that wire any smaller than an AN-6 gauge be used. You will notice from Fig. 10-1, that when considering replacing copper wire with aluminum you should use a wire having two wire *numbers* larger. For example, a number four copper wire will carry 135 amperes, and it will take a number two aluminum wire to carry as much current.

When choosing a wire to be used in an installation, you must consider the amount of current the wire will carry before it heats up enough to cause the insulation to smoke and also the voltage drop caused by the resistance of the wire. Fig. 10-2 is the wire chart familiar to most A & P technicians. With the use of this chart you can determine the wire size needed for an installation and you can check both its current-carrying capability and the voltage drop.

Let's assume that you want to install a cabin air conditioning system that requires 60 amperes, and is a continuous load. The wire will be installed as a single wire, that is, it is not in a bundle or a conduit, and the airplane has a 28-volt electrical system. The wire must be 30 feet long to connect the bus with the load.

First, locate the diagonal line representing 60 amperes, and follow it until it intersects the horizontal line for 30 feet in the 28-volt column. These lines intersect between the vertical lines representing six gauge and eight gauge, and the intersection is between curves 1 and 2. This means that a six-gauge copper wire is sufficient to carry the load for the required distance, and while the current is excessive for installation in a bundle or conduit, it is permissible to use it in free air. If it were necessary to run our wire in a bundle, we would have to back up the 60-ampere diagonal line to the first wire size that would get us above curve 1. We find that we can use a four-gauge wire; it would carry our load without excessive heat for the bundle.

You will notice that the wire chart of Fig. 10-2 has four columns for the circuit voltages. For wiring installation in a 12 (or 14) volt system it is permissible to have only a half-volt drop for equipment that is operated continuously. If it is operated intermittently, a one-volt drop is permissible. In a 24, or 28 volt system, a one-volt drop is permitted for a continuous load and two volts for an intermittent load. Higher voltage systems allow a larger voltage drop. A 115-volt system allows a drop of four volts continuous and eight volts intermittent, and a 200-volt system can tolerate seven volts continuous and 14 volts intermittent.

2. Wire Terminals

Wires that are to be installed on a terminal strip, such as the one in Fig. 10-3, are terminated with pre-insulated crimp-on terminals. These terminals are color coded to identify the wire size with which they are used. Red terminals are used with wire sizes 22 through 18, blue fits 16 and 14 gauge wire, and yellow terminals are to be used on 12 and 10 gauge wire. Note: Never stack more than four terminals on one stud.

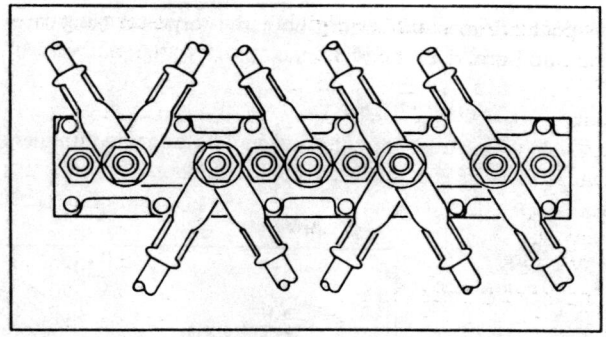

Figure 10-3.
Proper wire
installation on a
terminal strip.

Figure 10-4.
Proper installation of
a solderless wire
terminal.

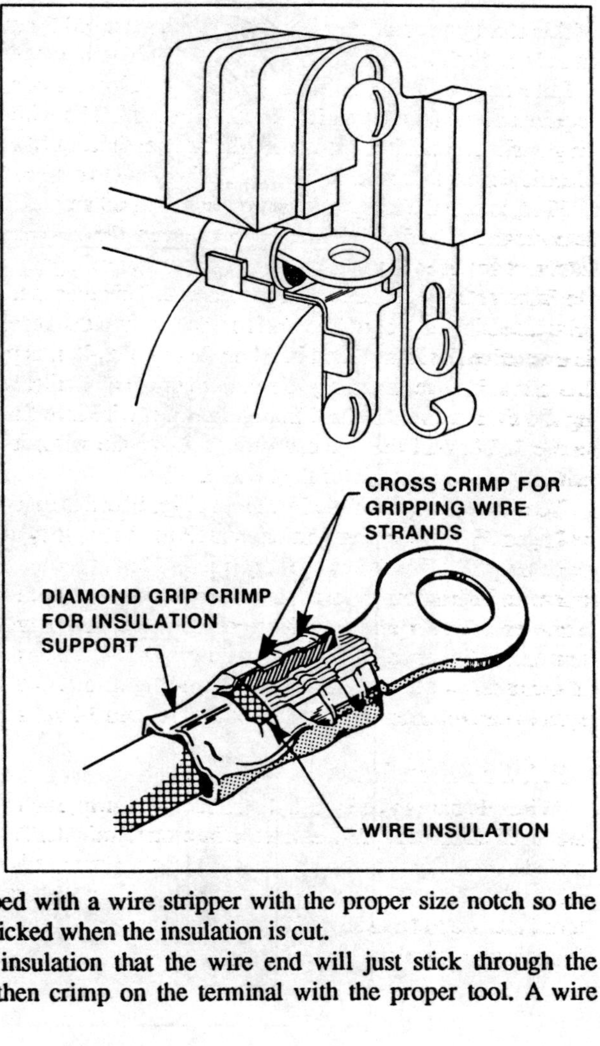

CROSS CRIMP FOR
GRIPPING WIRE
STRANDS

DIAMOND GRIP CRIMP
FOR INSULATION
SUPPORT

WIRE INSULATION

    The wire is stripped with a wire stripper with the proper size notch so the strands will not be nicked when the insulation is cut.

    Remove enough insulation that the wire end will just stick through the terminal body and then crimp on the terminal with the proper tool. A wire terminal crimper will crimp both the terminal body over the wire and the insulation grip over the wire insulation.

3. Connector Plugs

When wires must be connected and disconnected frequently, connector plugs similar to those seen in Fig. 10-5 are used.

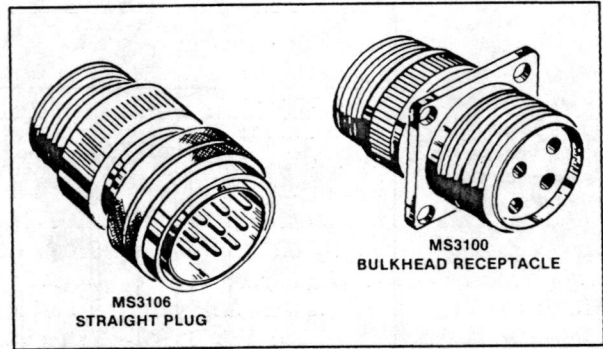

Figure 10-5.
Typical connector plugs.

MS3100
BULKHEAD RECEPTACLE

MS3106
STRAIGHT PLUG

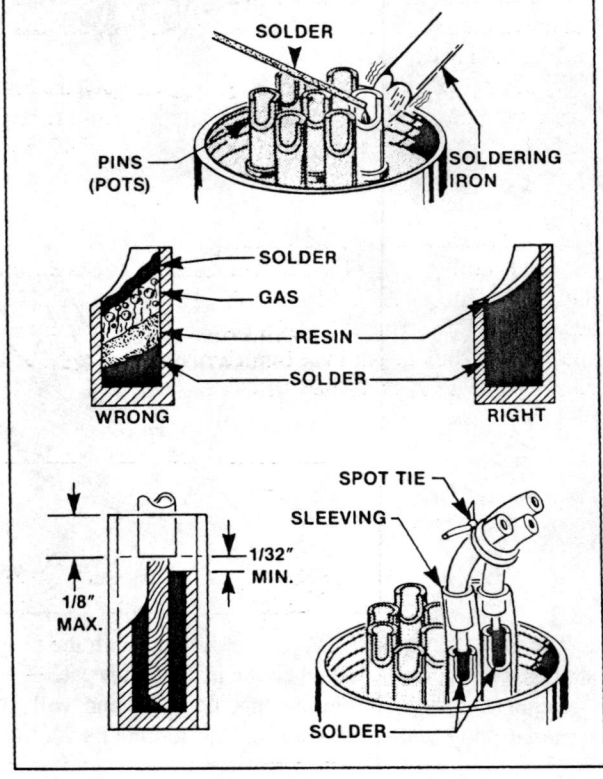

Figure 10-6.
Proper method of soldering a wire into the pot of a connector plug.

SOLDER

PINS
(POTS)

SOLDERING
IRON

SOLDER

GAS

RESIN

SOLDER

WRONG

RIGHT

1/32"
MIN.

1/8"
MAX.

SPOT TIE

SLEEVING

SOLDER

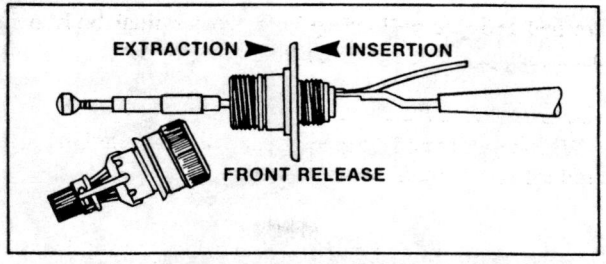

Figure 10-7.
Insertion/extraction of pin and socket connectors.

These plugs may have the wires soldered into the pots, or special pins or sockets may be crimped onto the wires and then pulled or pushed into the plug. Fig. 10-6 illustrates the correct way to solder a wire into a plug.

The wire is stripped so it will reach to the bottom of the pot and have about 1/32 inch of bare wire sticking out of the top. Slide a short piece of polyvinylchloride tubing, often called "spaghetti," over the wire and heat the outside of the pot with the proper size soldering iron. Fill the pot almost full of the correct solder, normally 60/40, resin core. CAUTION: *Do not use any acid-core solder for soldering any electrical equipment.* Slip the bare end of the wire into the molten solder, and when it is firmly on the bottom, hold it still and remove the iron. As soon as the solder loses its gloss, it is hard enough for you to release the wire. The solder should flow up around the lip of the pot, but not run down the side, nor should it wick up the wire and make it stiff between the pot and the insulation.

A newer generation of connector plugs, designed for faster and more efficient production, has its pins and sockets crimped into the wires and then they are pushed or pulled into the plug by means of a special insertion tool (Fig. 10-7).

C. Bundling And Routing

Wires coming from either a terminal strip or a connector plug should all be laid parallel and tied in a bundle. One of the methods finding a great deal of acceptance today is the use of a patented plastic strap called a TY-RAP. The strap is wrapped around the bundle and pulled tight with a special tool that cinches up the strap and cuts off its end.

Figure 10-8.
Individual spot ties for bundling electric wires.

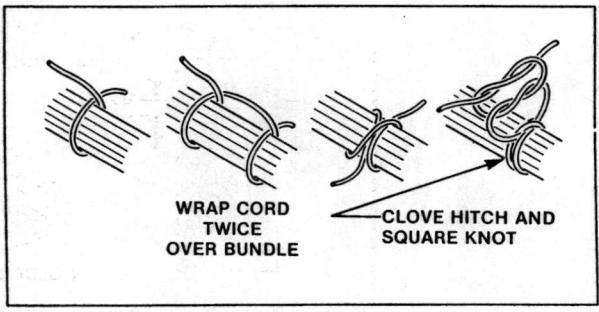

WRAP CORD
TWICE
OVER BUNDLE

CLOVE HITCH AND
SQUARE KNOT

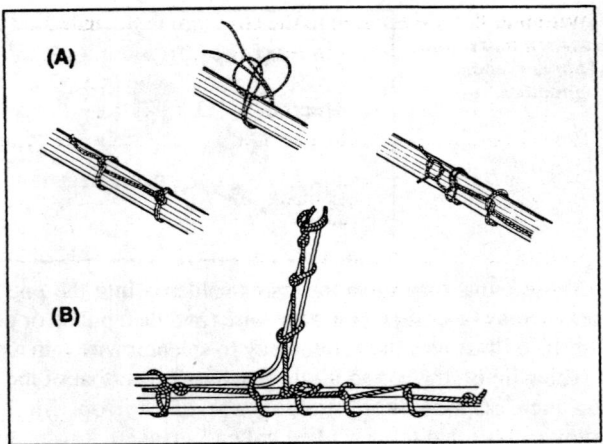

Figure 10-9.
Wire bundle laced
with waxed string.

(A)

(B)

Wire bundles may also be tied with spot ties along the run of the bundle. About every couple of inches two half-hitches are thrown around the bundle with waxed nylon or linen string, the ends pulled tight and secured with a square knot (Fig. 10-8). A faster method of securing wire bundles is to lace them with waxed string. Fig. 10-9 shows the method used. A bowline on the bight is used to start the lacing and single half-hitches are used along the run of the bundle. The lacing is terminated with a clove hitch (two half-hitches) secured with a square knot. Fig. 10-9A shows the way double string lacing is done, and the way part of a bundle is branched off from the main run of wire (Fig. 10-9B).

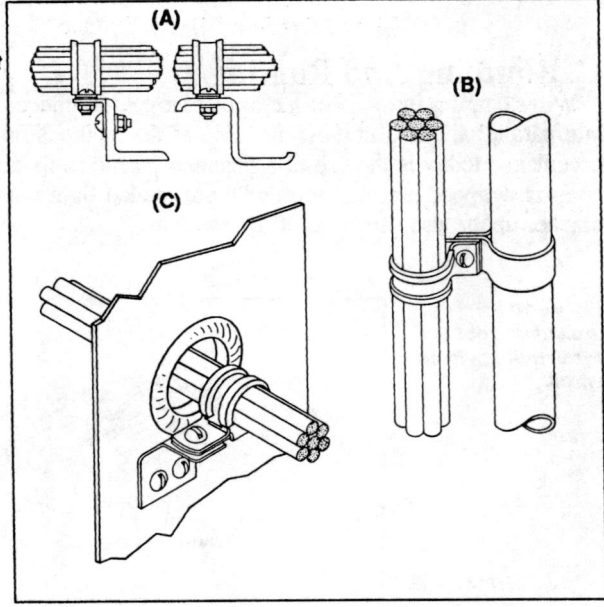

Figure 10-10.
Wire bundle
attachment to aircraft
structure.

(A)

(B)

(C)

Wire bundles are secured to the structure with cushioned clamps, attached in ways similar to that shown in Fig. 10-10.

Figure 10-10A: Attachment to sheet metal structural members.

Figure 10-10B: Attachment to tubular structure.

Figure 10-10C: Support of bundle passing through a hole in a bulkhead or frame.

Any time the bundle goes through a bulkhead or a frame, it must be centered in the hole, and if there is any possibility of chafing, the edges of the hole must be protected with a grommet. When there is a long run of bundled wire, it must have no more than a half inch of slack between the adjacent supports (Fig. 10-11). Be sure that the wire bundles do not pass through areas where the temperature is high, and if they must be routed parallel to fuel lines, be sure the wire bundle is above the line and never supported from it.

Figure 10-11.
Long run bundle,
with half-inch of
slack between
supports.

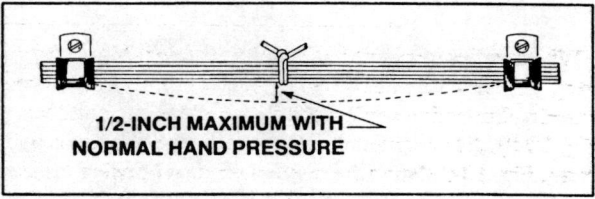

1/2-INCH MAXIMUM WITH
NORMAL HAND PRESSURE

QUESTIONS:

1. *What is meant by a trip-free circuit breaker?*

2. *Are automatic reset circuit breakers approved for use in aircraft circuits?*

3. *What size aluminum wire could be used to replace a two-gauge copper wire?*

4. *What would be the smallest wire you could use to carry a continuous 50-ampere load for 10 feet in a 14-volt airplane? The wire will be run in free air; that is, not in a bundle or conduit.*

5. *What color pre-insulated terminal would be used on a 12-gauge wire?*

6. *If a wire bundle is run parallel with a fuel line, which should be above the other?*

Glossary

This glossary of terms is provided to serve as a ready reference for the word with which you may not be familiar. These definitions may differ from those of standard dictionaries, but are in keeping with shop usage.

Accelerator winding   A series winding used on vibrating-type voltage regulators, which, when the points open, decreases the magnetic field immediately, allowing the points to close more rapidly.

Alternator   A special type of generator in which alternating current electricity is generated in the fixed windings and is converted to direct current by solid-state rectifying diodes.

Ammeter   An electron flow meter, used to measure electron flow in amperes.

Armature   The rotating element of an aircraft generator. Load current is generated in the armature.

Armature reaction   The distortion of the generator field flux by the current flowing in the windings of the armature.

Base   The electrode in a transistor to which the signal is applied.

Battery   A device made up of a number of individual cells, used to store electricity by converting it into chemical energy. Electrons are caused to flow from one pole, the anode, to another pole, the cathode, by a chemically produced potential difference.

Bomb tester   A spark plug tester in which the plug is exposed to approximately 200 psi of air pressure, and high voltage is applied to the center electrode. Plugs which can spark in this atmosphere are considered to be acceptable.

Bootstrapping   The technique with which something is brought into the desired state through its own action.

Breaker points   Interrupter contacts in the primary circuit of a magneto or battery ignition system. They are opened by a cam the instant the highest current flows in the primary circuit, thus producing the maximum rate of collapse of the primary field.

Cam   An eccentric plate or shaft used to impart motion to a follower riding on its surface or edge.

Cam, compensated   The magneto cam used on high performance radial engines. One lobe is provided for each cylinder, and the lobes are ground in such a way that the magneto points will open when the piston is a given linear distance from the top of the cylinder, rather than a given angular distance. This compensates for the relationship of the master rod pistons and those connected to the crankshaft through the link rods.

Ceramic   A clay-like material composed primarily of magnesium and aluminum oxide, which may be molded and fired to produce an excellent insulating material.

Chamfered tooth   The tooth of a gear on a rotating magnet or distributor gear, which is beveled to identify it for use when timing a magneto.

Circuit breaker   A circuit protection device which will open a circuit having an excessive flow of current. Circuit breakers may normally be reset in flight.

Circuit breaker, automatic-reset   A circuit protection device which will open the circuit when a current overload occurs and which, when the overload is no longer present, will reset itself automatically. Automatic-reset circuit breakers are not approved for use in aircraft.

Circuit breaker, trip-free   A circuit protection device which will open a circuit when a current overload exists, regardless of the position of the control handle.

Circular mil   A measurement of area equal to that of a circle having a diameter of one one-thousandth of an inch.

Clove hitch   Two half-hitches tied around aircraft wiring bundles to hold them.

Coil   A conductor consisting of turns of wire in which the magnetic field around one turn cuts across the other turns, increasing the inductive effect of the wire.

Collector   The electrode through which conventional current leaves a transistor.

Commutator   The copper bars on the end of a generator armature to which the rotating coils are attached. Alternating current is generated in this armature, and brushes riding on the commutator act as a mechanical switch to convert it into direct current.

Compensating winding   A series winding in a high output generator, wound between the main pole and the interpoles to aid in brushless commutation.

Conductor   A material whose outer ring electrons are loosely bound. A relatively low voltage will cause a flow of these electrons.

CSD   Constant speed drive.   A unit used to maintain a constant rotational speed of aircraft AC alternators even though the engine speed varies.

Current   Throughout this book "current" refers to electron current, from negative to positive in the external circuit. Conventional current, or "Franklin current," is said to flow in the opposite direction.

Current, alternating   A flow of electrons which constantly changes, periodically reversing its direction.

Current, conventional   Current flowing in an electrical circuit from positive to negative, outside the power source.

Current, direct   A flow of electrons in one direction throughout a circuit.

Current, eddy   Current induced into the core of a coil or transformer by current flow in the winding. Eddy currents cause power losses and are minimized by laminating the iron cores.

Current limiter   A device which limits generator output to a level within that rated by the generator manufacturer.

Current, primary   The alternating or pulsating current which flows in the primary winding of a transformer and induces a current into the secondary winding.

Current, pulsating direct   Direct current which has been chopped by a vibrator or chopper, and which changes from zero to maximum, then back to zero. This produces the changing current required for use in a transformer.

Current, secondary   The current that flows in the secondary winding of a transformer.

Delta winding   The connection of a three-phase AC output in which the ends of all three phases are joined to form a triangle or a delta.

Diode   A two-element electron control valve.

Distributor   That part of a high-tension magneto which distributes high voltage to each spark plug at the proper time. Distributors for low-tension ignition systems distribute low voltage to the transformers at each plug, at their proper time.

DVRCR   Differential-voltage reverse-current relay. A form of reverse-current relay used for high output systems, in which the relay closes when the generator voltage is at a specific value above the voltage of the battery.

E-gap   The number of degrees of magnet rotation beyond its neutral position at which the primary breaker points, in a magneto, open. It is at this point that the primary current flow is the greatest, and therefore the rate of collapse of the primary field will induce the greatest voltage into the secondary winding.

Electron flow   The current flow in a circuit, which is actually the flow of electrons. Electrons flow from negative to positive in the external circuit.

Element   One of the basic known chemical substances that cannot be divided into simpler substances by chemical means.

Emitter   The electrode of a transistor through which conventional current enters.

Engine, four-cycle   The most common event cycle for aircraft engines. The four-stroke, five-event cycle consists of an intake stroke, in which the piston moves inward with the intake valve open, a compression stroke in which the piston moves outward with both valves closed. Near the top of the compression stroke, ignition occurs. The power stroke is an inward stroke of the piston with both valves closed, and the exhaust stroke occurs when the piston moves outward with the exhaust valve open. At this point the cycle begins again.

Engine, radial   An engine, popular from the time of World War I through World War II, in which the cylinders radiated out from the central crankshaft. Single-row radial engines have odd numbers of cylinders, and more than one row of cylinders may be stacked to form two- or four-row radial engines.

Engine, rotary   A radial engine, popular in the first World War. Its cylinders rotated around the fixed pistons, and the propeller was attached to the rotating case.

Engine, two-cycle   A reciprocating engine in which a power impulse occurs on each stroke of the piston. As the piston moves outward, fuel-air mixture is drawn into the crankcase, below the piston, while above the piston the mixture is compressed. Near the top of the stroke, ignition occurs and, as the piston moves downward, power is exerted on the crankshaft. Near the bottom of the stroke, exhaust action takes place on one side of the cylinder and intake action occurs on the opposite side.

Field excitation   Direct current supplied to the field of an alternator or a generator to produce magnetic flux which is cut by the conductors in the armature or stator.

Firing position   The position of a piston in the cylinder of a reciprocating engine at which time ignition should occur. Igniting the mixture at this position allows peak cylinder pressure to occur shortly after the piston passes top center.

Flux, magnetic Invisible lines of magnetic force that are assumed to exist between the poles of a magnet. Traditionally they are given the direction from north pole to south pole. One line of magnetic force is called one maxwell.

Flux, resultant The concentration of lines of flux in the frame of a magneto due to the primary flux of the rotating magnet and the flux caused by the flow of primary current. The field from the primary current sustains the flux beyond the point the magnet passes through the neutral position.

Flux, static The concentration of lines of flux in the frame of a magneto due to the rotation of the rotating magnet. At full register, the lines of flux are maximum, and at the neutral positions, the lines are minimum.

Forward bias The arrangement of voltage polarities for a transistor or a diode which allows conduction.

Full register position The position of the rotating magnet in a magneto when the poles are fully aligned with the pole shoes of the magneto frame. At this point the maximum number of lines of flux flow in the frame.

Fuse A circuit protection device that will open the circuit by melting a fuseable link when an excess of current flows.

Generator A mechanical device consisting of a conductor being turned within a magnetic field, used to produce electricity by electromagnetic induction.

Generator, compound-wound A generator which has both a series and a shunt field.

Generator, series-wound A generator in which the field and armature are in series.

Generator, shunt-wound A generator in which the field and armature are in parallel.

Growler Test equipment used to check generator and starter armature for shorts. The growler forms the primary of a transformer and the armature forms the secondary. Shorts show up as they cause vibration of a piece of metal—such as a hacksaw blade—held over the armature.

Heat A form of energy associated with the motion of molecules within a material.

Heat sink A device on which semiconductors may be mounted to absorb the heat that would normally tend to damage them.

Hole A vacancy in the valence structure of an element which will accept an electron from an outside source.

Horsepower HP. The standard unit of power used for mechanical measurement. It is equal to 33,000 foot-pounds of work done in one minute. Electrically, it is equal to 746 watts.

Impulse coupling A spring-loaded coupling between a magneto and its drive gear. When the engine is being turned over slowly, the magnet is restrained by stops, and the spring is wound. At the proper time for the starting spark to occur, the spring is released and the magnet is spun, producing a hot, late spark. When the engine starts, centrifugal force holds the coupling engaged so that it acts as a solid unit.

Inductance That property of a conductor which causes an electromotive force, or voltage, to be generated when lines of magnetic force cut across it.

Induction motor, capacitor-start An AC motor whose rotor is excited by voltage induced from the field windings. A second winding whose phase is shifted by a capacitor is used to provide a rotating field for starting. When the motor gets up to speed, a centrifugal switch opens the circuit in which the capacitor is situated.

Induction vibrator A coil and set of contact points which produce pulsating direct current from straight direct current. Pulsating DC may be used in the primary winding of a magneto to produce a high voltage in the secondary winding.

Internal timing The timing of the relationship between the E-gap position of the rotating magnet and the opening of the breaker points.

Interpoles Field poles placed between regular generator fields. The windings around them are in series with the armature. Interpoles are used to prevent arcing at the brushes caused by armature reaction.

Iridium An extremely hard and brittle metallic element of the platinum group, which is used for electrodes of fine-wire spark plugs which must operate in extremely severe lead conditions.

Junction The point at which two conductors or circuits join.

KVAR Kilovolt amperes, reactive. A measure of reactive power.

Laminated core   The core of a coil or transformer, consisting of a stack of thin, soft iron sheets, insulated from each other by an oxide film or varnish. Laminated cores minimize eddy current losses.

Left-hand rule for generators   If the fingers of the left hand encircle a conductor in the direction of the lines of magnetic flux, the thumb will point in the direction of the electron flow.

Magnet   A device or a material that has the property of attracting or repelling magnetic materials. Lines of magnetic flux link its external poles, and a conductor cutting across the flux will have a voltage induced in it.

Magnet, permanent   A ferrous metal, or alloy of ferrous metals, usually containing nickel and cobalt, in which the magnetic domains are aligned and tend to remain aligned. Lines of magnetic flux join the poles of the permanent magnet so that an electrical current may be generated when these lines of flux are cut by a conductor.

Magneto, booster   A small high-voltage magneto, usually turned by hand, used to produce a hot spark for starting reciprocating engines. The output for the booster magneto is fed into a trailing finger on the distributor which fires the cylinder following the one in position for normal ignition.

Magneto, dual   Actually a single magneto housing which holds one rotating permanent magnet and one cam, with two sets of breaker points, two condensers, two coils, and two distributors. For all practical purposes, this constitutes two ignition systems.

Magneto, high-tension   A magneto which consists of a rotating magnet, cam, breaker points, condenser, and a coil with a primary and a secondary winding. The output of the secondary winding goes to a distributor, then to the spark plugs.

Magneto, low-tension   A magneto consisting of a rotating magnet, a cam, breaker points, condenser, a coil with only the primary winding, and a carbon-brush-type distributor. The primary current is directed through the distributor to a coil for each individual spark plug. These coils have a primary and a secondary winding. The high voltage is generated at the spark plug.

Magneto, Scintilla   The trade name of a Swiss designed and built magneto. The forerunner of the current Bendix magnetos.

Motor, shaded-pole   A low torque AC induction motor whose rotating field is provided by the inductive action of shading poles on diametrically opposed pole pieces.

111

Motor, universal   A series-wound motor that will operate on either alternating or direct current.

Neutral plane   An imaginary line drawn perpendicular to the resultant flux in a generator. For arcless commutation, the neutral plane should lie directly over the plane of the brushes.

Neutral position   The position of the rotating magnet of a magneto between the pole shoes. In the neutral position no lines of flux flow in the magneto frame.

Parallel   Having more than one path for electron flow from one side of the electrical source to the other.

Paralleling   Controlling the output of more than one generator so that they will equally share the load.

Platinum   A hard, gray metallic element with an extremely high melting point. It is used for the electrodes of fine-wire spark plugs.

Polarity   The property, of an electrical device, having two different types of electrical charge: positive (deficiency of electrons) or negative (excess of electrons).

Pulsate   To expand and contract rhythmically, yet not change direction.

Reach   The length of the shell thread of a spark plug. For 18-mm spark plugs, long-reach plugs are threaded for $13/16$ of an inch and short-reach plugs, for $1/2$ inch.

Reactive power   The product in an AC circuit of the total voltage and the total current.

Real power   The power in an alternating current circuit which is the product of the voltage and the current in phase with the voltage. It is the voltage times the current times the power factor.

Resistor   An electrical circuit element used to provide a voltage drop by dissipating some of the electrical energy in the form of heat.

Retard points   An auxiliary set of breaker points in a magneto equipped with the Shower of Sparks starting system. These points are operative only during the starting cycle and open later than the run, or normal, points. This provides a late, or retarded, spark.

Reverse bias The arrangement of voltage polarities of a transistor or a diode which does not allow conduction.

Rotor The rotating element in an alternator. It is excited by direct current, and the interlacing fingers on the two faces of the rotor form the alternating north and south field poles.

RPM Revolutions per minute. Rotational speed of an engine crankshaft.

Self-excited Field excitation coming from the generator output without being regulated.

Semiconductor A device or material that will conduct under some conditions and act as an insulator under others.

Short circuit A path for electrons to flow from one level of potential to another, without completing a useful circuit.

Silicon An element, normally a resistor, used in the manufacture of semiconductor devices.

Silicon glaze A shiny, brown, glass-like deposit on the nose insulator of a spark plug that has been operated in sandy or dusty conditions. This glaze is an insulator at low temperatures, but at high temperatures it becomes conductive.

Silicon, N-type Silicon which has been doped with an impurity having five valence electrons.

Silicon, P-type Silicon doped by an impurity having three valence electrons.

Sine wave The wave-form of alternating current produced by a rotary generator. Its amplitude at any time is proportional to the sine of the angle through which the generator has turned.

Slip ring A smooth circular ring used to put field current into a DC alternator.

Spaghetti An insulating tubing slipped over wires.

Spark plug, all-weather A shielded spark plug in which the ceramic is recessed into the shell so a resilient grommet on the harness lead can provide a watertight seal. All-weather spark plugs are easily identified by three-quarter-inch, twenty-thread-per-inch shielding.

Spark plug, cold   A spark plug in which the nose insulator provides a short path for heat to travel from the center electrode to the shell. Cold spark plugs are used in high compression engines to minimize the danger of pre-ignition.

Spark plug, fine-wire   A spark plug using platinum or iridium electrodes. The small electrodes allow the firing end cavity to be open to provide better scavenging of lead oxides from the plug. Heat transfer characteristics of the fine wires prevent their overheating.

Spark plug, hot   A spark plug with a long nose insulator in which the heat transferring from the center electrode into the shell has a long path to travel. Hot spark plugs are used in engines which operate relatively cool, and they keep the center insulator hot enough to prevent accumulation of lead oxides.

Spark plug, massive-electrode   Spark plugs using two, three, or four large nickel-alloy ground electrodes.

Spark plug, resistor   A composition resistor installed in the barrel of most shielded spark plugs. The resistor limits the current which is stored in the capacitive effect of the shielding and minimizes electrode erosion.

Spark plug, shielded   A spark plug completely encased in a steel shell. The radiated energy from the spark is conducted to the ground through the shielding, preventing radio interference.

Thread chaser   A tool used to remove contamination from a threaded device.

Transistor   A semiconductor device having three electrodes, similar in use to the vacuum tube.

TY-RAP   A patented nylon strap used to hold wire bundles together.

VAR   Volt-ampere, reactive.

Voltage regulator, carbon pile   A regulator for direct current generators which varies resistance in the field by changing the resistance of the stack of carbon discs.

Voltage regulator, transistor   A voltage regulator for DC generators or alternators which uses a transistor to control the field current. A zener diode senses the voltage to be controlled.

Voltage regulator, transistorized   A voltage regulator for DC generators or alternators which uses a transistor to control the flow of field current, but vibrating points to sense voltage and control the transistor.

Voltage regulator, vibrating-type A voltage regulator for direct current generators or alternators which uses vibrating points to sense voltage and provide a varying resistance for the generator field current.

Voltmeter An electrical measuring instrument used to measure electrical pressure or voltage.

Y-winding The connections of a three-phase AC alternator in which one end of each of the three windings is common.

Zener diode A semiconductor device that allows free electron flow in one direction and restricts flow in the opposite direction until a specific voltage has been reached.

Aircraft Ignition And Electrical Power Systems

Final Examination

STUDENT _____

GRADE _____

Place a circle around the letter for the correct answer to each of the following questions.

1. What is meant by the E-gap angle of a magneto?

   A. The angular distance between the full register position of the magnet and the position at which the points open.

   B. The angular distance between the neutral position of the magnet and the position at which the points close.

   C. The angular distance between the neutral position of the magnet and the position at which the points open.

   D. The angular distance between the full register position of the magnet and the position at which the points close.

2. At which position of the rotating magnet does the maximum primary current flow in a magneto?

   A. At the full register position.

   B. In the neutral position.

   C. A few degrees before the neutral position.

   D. A few degrees after the neutral position.

3. How is a magneto with an impulse coupling timed to an engine?

   A. The magneto should be installed in such a way that the impulse coupling should snap just as the piston reaches its proper firing position.

B. The magneto should be installed in such a way that the impulse coupling should snap just as the piston reaches top center.

C. The engine should be rotated until the impulse coupling snaps, then backed up until the points just close. The points should just close at the proper firing position.

D. The engine should be rotated until the impulse coupling snaps, backed up until the points close, then moved forward until they break. They should break at the proper firing position.

4. Some magnetos have two sets of breaker points. What is the function of the second set?

A. They open later than the normal points and are used to provide the proper timing of the spark for starting.

B. They open earlier than the normal points and are used to provide the proper timing of the spark for starting.

C. They are back-up points to be used in the event of the failure of the normal points.

D. The only reason for having a second set of points would be to operate a tachometer.

5. What is one of the main advantages of a fine-wire spark plug over a massive-electrode spark plug?

A. Fine-wire spark plugs require a higher voltage to jump their gaps.

B. Fine-wire spark plugs are less susceptible to lead fouling than massive-electrode spark plugs.

C. Fine-wire spark plugs have a much longer service life than massive-electrode spark plugs.

D. Fine-wire spark plugs cannot be fouled with tetraethyl lead deposits.

6. What is meant by the term hot spark plug?

A. One that burns your fingers when you take it out of the engine.

B. A spark plug with a long nose insulator—one which provides a long path for the heat to travel before it is dissipated into the shell.

126

C. A spark plug with a short nose insulator—one which provides a short path for the heat to travel before it is dissipated into the shell.

D. That classification of spark plug which should be used in a high powered, high-compression engine—a hot engine.

7. What is the function of a resistor in a resistor spark plug?

A. It is used to reduce radio interference.

B. It prevents the spark plug from sparking until an extremely high voltage is reached in the magneto secondary.

C. It serves as a lock to prevent the spark plug from unscrewing from the hole if it has not been put in tight enough.

D. It minimizes electrode erosion by limiting the peak current that is allowed to flow.

8. Which one of these is a recommended way of cleaning a spark plug?

A. Remove most of the lead deposits with a vibrator-type cleaner and very lightly abrasive. Blast the firing end cavity with a glass bead or aluminum oxide abrasive.

B. Soak the entire spark plug in a decarbonizing solution, and then remove the residue with an abrasive blast.

C. Remove all of the contaminants from the firing end of the spark plug with a sand blaster, using pure silica sand.

D. Blast all of the lead deposits from the firing end cavity with aluminum oxide abrasive.

9. What is true about gapping spark plugs?

A. The ground electrode of a fine-wire spark plug is brittle and should not be moved more than just enough to close the gap.

B. The ground electrode of a fine-wire spark plug can be moved with very little danger of its breaking, as it is made of either platinum or iridium, which are fairly ductile metals.

C. It is necessary to adjust the gap of only one ground electrode on a massive-electrode spark plug having three ground electrodes. The other two are there only as spares.

D. If a ground electrode on a fine-wire spark plug is moved over too much, you can open the gap by very carefully prying against the center electrode with a small screwdriver.

10. What is the danger of installing a spark plug with too little torque?

A. None. If you use only about half the recommended torque, your spark plug may last longer because you will not stretch the threads.

B. The spark plug electrode may not stick far enough into the cylinder to ignite the mixture if it is not torqued in tight enough.

C. Too little torque could provide an inadequate seal, and hot gases could escape around the threads, damaging both the spark plug and the cylinder head.

D. The spark plug may vibrate in its hole and wear the threads in the cylinder head.

11. Which is **not** one of the factors which determines the amount of voltage produced in an electromagnetic generator?

A. The number of lines of flux.

B. The number of conductors cutting across the flux.

C. The size of the conductor cutting across the flux.

D. The speed at which the conductor cuts across the flux.

12. How does armature reaction cause brush arcing?

A. It decreases the spring pressure holding the brush onto the commutator.

B. It moves the neutral plane away from the brush location.

C. It increases the output voltage of the generator.

D. It decreases the current in the armature.

13. Why do some high-output generators have interpoles and compensating windings?

    A. To increase the armature reaction.

    B. To increase the voltage output of the generator.

    C. To decrease the arcing at the brushes.

    D. To operate the generator at a constant speed.

14. How does a vibrator-type current limiter decrease the current output of a generator?

    A. It places a resistor in the field circuit, decreasing the output voltage.

    B. It places a resistor in the armature circuit, decreasing the armature current.

    C. It opens the load circuit when the current is excessive.

    D. It grounds the field circuit when the load current becomes excessive.

15. What current flows through the voltage regulator points on a transistorized voltage regulator?

    A. The load current.

    B. The field current.

    C. The transistor emitter current.

    D. The transistor base current.

16. What does a DC alternator use in place of the brushes and commutator used on a DC generator?

    A. Brushes and slip rings.

    B. Semiconductor diodes.

    C. Transistors.

    D. Permanent magnet rotor.

17. Which transistor circuit would have full alternator field current flowing?

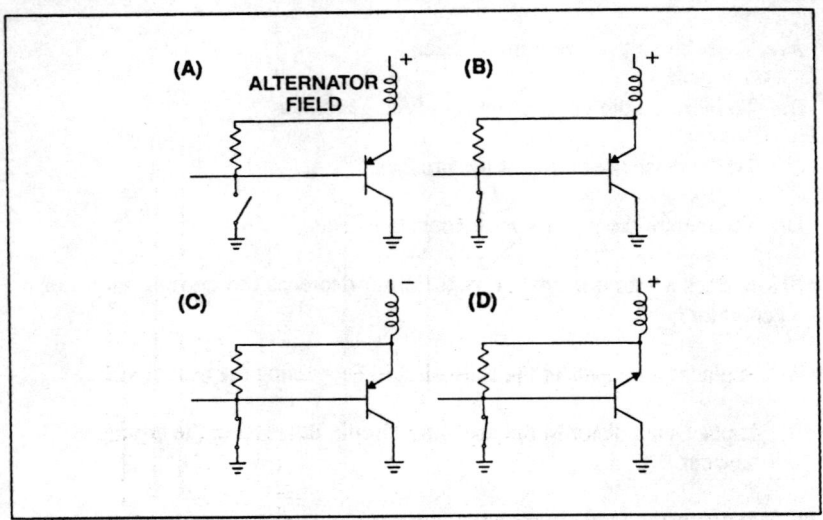

18. What is used to check a generator armature for shorts?

    A.   A growler.

    B.   A high-range ohmmeter.

    C.   A voltmeter and an ammeter.

    D.   An oscilloscope.

19. What is used to test the semiconductor diodes in a DC alternator?

    A.   An AC test light.

    B.   A high-range ohmmeter.

    C.   A low-voltage ohmmeter.

    D.   An AC voltmeter.

20. What size wire would be proper to use to carry 100 amperes to a starter motor? The airplane has a 14-volt electrical system, the wire must be six feet long and it is not to be routed in a bundle. Starters are intermittent loads. (Refer to wire chart.)

A. 2 gauge.

B. 4 gauge.

C. 6 gauge.

D. 8 gauge.

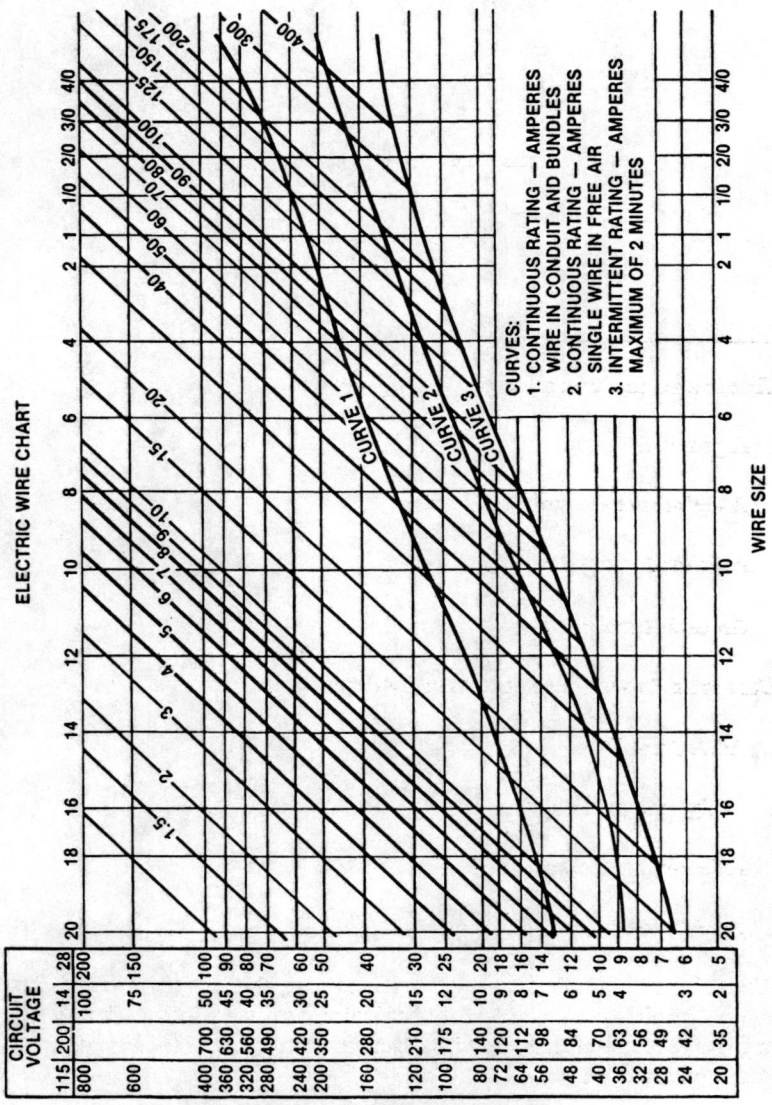

ELECTRIC WIRE CHART

CURVES:
1. CONTINUOUS RATING — AMPERES
   WIRE IN CONDUIT AND BUNDLES
2. CONTINUOUS RATING — AMPERES
   SINGLE WIRE IN FREE AIR
3. INTERMITTENT RATING — AMPERES
   MAXIMUM OF 2 MINUTES

WIRE SIZE

WIRE LENGTH IN FEET PER VOLT DROP

Aircraft Ignition And Electrical Power Systems

Answers To Final Examination

1.	C	11.	C
2.	D	12.	B
3.	D	13.	C
4.	A	14.	A
5.	B	15.	D
6.	B	16.	B
7.	D	17.	B
8.	A	18.	A
9.	A	19.	C
10.	C	20.	B